DICTIO
SCIENCE
FOR EVERYONE

BLOOMSBURY
DICTIONARY OF
SCIENCE
FOR EVERYONE

HERMAN SCHNEIDER & LEO SCHNEIDER

Jacket picture photographed by Ruth Bayer with kind permission of the laboratory of Molecular Biology (Department of Crystallography), Birkbeck College, London.

All rights reserved; no part of this publication may be reproduced, stored in a retrieval system, or transmitted in any form or by any means, electronic, mechanical, photocopying or otherwise, without the prior written permission of the Publishers.

Copyright © 1988 by Herman Schneider and Leo Schneider.
Additional material © Bloomsbury Publishing Limited.

First published in Great Britain, 1989 by Bloomsbury Publishing Limited, 2 Soho Square, London W1V 5DE.

This paperback edition published 1990.

10 9 8 7 6 5 4 3 2 1

British Library Cataloguing in Publication Data

Bloomsbury dictionary of science for everyone
 1. Science — Encyclopedias
 I. Schneider, Leo II. Schneider, Herman
 503'.21

ISBN 0 7475 0678 7

Designed by Panic Station
Printed in Great Britain by Cox and Wyman Ltd, Reading, Berkshire.

Preface

This book is intended for two types of readers:

1. Scientists
2. Non-scientists

Type 1 is exemplified by a renowned geologist friend who frankly admits that all he knows about DNA is that it has to do with heredity and life, that Watson and Crick got the Nobel Prize for it — and that the details are beyond him unless he invests more time and trouble than he can spare.

Type 2 is exemplified by an exasperated neighbour who phoned recently (of which more in a moment). In general, non-scientists find that even when they don't exactly seek out science, science seeks them out, in a variety of media and messages:

- The TV weatherman points regretfully to an *occluded front* on the map. What's a front? How is it occluded? Why does he regret it? What will it do to your plans for a day at the beach?
- A newspaper columnist surmises that the power of OPEC may be broken by the power from *tokamak*. OPEC — that's clear, but tokamak? Will it give you more miles per gallon, or what?
- The youthful-looking lady in the full-page, full-colour magazine ad coyly confesses she's a grandmother; she owes her baby-skin complexion to the scientific catalytic hormone formula in Preparation Q Scientific? *Catalytic? Hormone?* Will it give *you* a baby complexion? Do you want a baby complexion?
- The obituary section reports the career of a *microbiologist* internationally known for research in *ribonucleic acid, deoxyribonucleic acid,* and *genetic mutations.* The genetic item arouses a sense of fuzzy familiarity, enough to make you wish you understood the other words.

You can't escape science — nor probably do you want to. You just wish its language would come into focus a bit more sharply. There's that fuzzy area around *nuclear.* Is it a threat? Is it related to the nucleus of a cancer cell? A cheap pocket calculator is described as containing over 20,000 electronic *transistors.* But isn't a transistor a gadget that emits screechy music and incomprehensible news reports? Is electronic the same as electric?

Of course, there's the traditional way out of this puzzlement. Look it up in the dictionary, encyclopedia, or similar reference book. That, in fact, was the motivating force behind this book; it originated in a spirit of wholesome, creative exasperation.

The aforesaid neighbour phoned, reporting on the search for a definition of *entropy*. The word had appeared in a news article about sunspots, lightly explained. Seeking a bit more depth, he found the following definitions:

In his standard dictionary:

> **en.tro.py** (ĕn'trə-pē) *n.* 1. A measure of the capacity of a system to undergo spontaneous change, thermodynamically specified by the relationship $dS = dQ/T$, where dS is an infinitesimal change in the measure for a system absorbing an infinitesimal quantity of heat dQ at absolute temperature T. 2. A measure of the randomness, disorder, or chaos in a system specified in statistical mechanics by the relationship $S = k \ln P + c$, where S is the value of the measure for a system in a given state, P is the probability of occurrence of that state, and k is a fixed and c an arbitrary constant. [German *Entropie*: Greek *en-*, in + *tropē*, a turning, change (see trop-[2] in Appendix*).]

In a science dictionary:

> **entropy**, n. [Phys.]. In a THERMODYNAMIC system the measure of the amount of heat ENERGY in it which it is not possible to make use of by changing it into MECHANICAL WORK, given as the RATIO of the amount of heat present to the TEMPERATURE.

In a one-volume encyclopedia:

> **ENTROPY**, en'trə-pē, is a measure of disorder or randomness in a physical system or, in information theory, a measure related to the information content of a message. The term 'entropy' was coined by the German physicist Rudolf Clausius in 1865 to denote a thermodynamic function, which he had originally introduced in 1854, that tends to increase with time in all spontaneous natural processes. This property of entropy led Clausius to his famous paraphrase of the first and second laws of thermodynamics: 'The energy of the universe is constant. The entropy of the universe tends toward a maximum'.

In a 20-volume 'home encyclopedia':

> **Entropy** The *entropy* of a system is a measure of its disorder, and in any change affecting a closed system the entropy can only *increase*, as in the above example. This is another way of stating the Second Law (due originally to Clausius, who first introduced the idea of 'entropy' in 1865), and it is interesting because it defines the direction of TIME. Most of the laws of physics would be equally valid if time ran backwards: for instance, a film of a swinging pendulum would be indistinguishable from the same film run backwards, but if a process involving an entropy change occurs (such as the stopping of the pendulum by air resistance or friction) it is immediately obvious which way the film should be run.

PREFACE

In a book of science essays for the general reader:

> In Ludwig Bolzmann's formulation the relationship between entropy S and order is given by the expression $S = k \log W$, where k is the universal Bolzmann constant and W the probability of the prevailing configuration of the system. Where the probability of a given configuration is unity, there is no possibility of alternative configurations. Under the circumstances, the entropy of the system is of course zero (remembering that $\log 1 = 0$). Such is the case of a perfectly pure crystal at 0°K (absolute zero), where heat motion is at a standstill.

In a dictionary of technical terms:

> **en'tropy** *(Phys.).* A thermodynamic conception that, if a substance undergoing a reversible change takes in a quantity of heat dQ at temperature T, its *entropy* is increased by $\frac{dQ}{TQ}$

Now the purpose of this preface is not to beat horses, whether dead or alive, but to point out some cogent facts:

- The definitions of *entropy* are most comprehensible to the person least needful of them: the reader with a good science background.
- All the definitions are entirely correct, as any physicist would agree. Yet the emphasis and treatment are so varied as to give the effect of relating to different words — a kind of six-blind-men-and-an-elephant approach.
- All were written with the laudable goal of being impersonal and objective — 'scientific' — thus omitting mention of the profound significance of entropy, the most wide-reaching physical event in the universe.

So that's how this book was nudged into being. We selected more than 1000 common words of science, according to the following criteria:

- *Statistical importance.* Over a period of time, we tracked the frequency of occurrence of terms in a number of widely read magazines and newspapers, and in TV and radio broadcasts.
- *Portmanteau value.* Some words were chosen because they are good containers; they enabled us to develop a whole sequence of related terms with maximum efficiency and minimum grief. For example, the phrase *elementary particles,* while relatively low on the frequency list served as a fine basket for carrying protons, electrons, mesons, bosons, gluons, quarks, and numerous other tiny items currently hot in science. However, if all you want is a bit about quarks, *sans* historical development from Democritus on, just look among the *Q*'s.
- *Whim* (see ALPHANUMERIC).

PREFACE

- *Door-opening value.* Some science concepts are easy to understand in their basic form, yet they open doors to much more complex and sophisticated applications of the same principle. Thus the concept of *resonance* can be understood by swinging four sugar cubes from a thread. What we observe helps us to understand how a microwave oven cooks a potful of soup without heating the pot, how the turning of a knob selects a single radio programme out of the hundreds clamouring for admission, and how an astronomer determines that a certain star contains a specific percentage of iron in its atmosphere.

Now for a final word or two about this alphabetical list of word definitions, this dictionary. *It's a bargain!* Unlike most dictionaries, this one will not suffer from rapid obsolescence. Consider what has happened in recent years to the meanings of some non-science words: *soap, gay, snow, hash, coke.* Or look at the long-term turnaround in a word such as *rival,* which in the 14th century meant the fellow who shares the river (Latin, *rivalis*) with you. Then the river part dropped out, leaving only the sharing, as in *Hamlet:*

> If you do meet Horatio and Marcellus,
> The rivals of my watch, bid them make haste.

Then the sharing idea turnèd around to competing, as with two suitors, rivals for a lady's hand, who shared only a desire for the lady. Think of all those words in that expensive *Oxford English Dictionary*, changing meaning under your very eyes, nibbling away at your investment.

But never in science! (Well, hardly ever.) In science, a rose is a rose is a rose — or to put it more scientifically, a flower of the family Rosaceae is a flower of the family Rosaceae is a flower of the family Rosaceae. And a quark is a quark is a quark — anywhere, anytime, under any political system. So this little book should retain its value, on and on and on, a continuing bargain.

At least until the next edition.

Note
'billion' is used throughout to denote 'a million millions' and 'trillion' 'a million million millions'.

ABLATION See SPACE TRAVEL

ABSOLUTE MAGNITUDE See MAGNITUDE

ABSOLUTE ZERO

The coldest possible temperature. It has never been achieved, although physicists have come within a few millionths of a degree. That ultimate bottom temperature is − 273.15°C (−459.7 °F).

Temperature is a measure of the agitation of the molecules of matter: much agitation = higher temperature; less agitation = lower temperature; virtually no agitation = lowest temperature.

In 1848 Lord Kelvin, a Scottish physicist, proposed a temperature scale that avoided negative numbers, such as '40 below zero', or '−40 degrees'. On Kelvin's scale the lowest possible temperature was labelled the zero point, hence 'absolute zero'. One unit, or degree, on this scale, is called a *kelvin,* and it is equal to one degree on the Celsius scale. Thus the freezing point of water, 0° Celsius (abbreviated to C), is 273 kelvins (abbreviated to K). The boiling point of water, 100°C, is 373 K.

AC See DC AND AC; RADIO

ACETYLCHOLINE See ALZHEIMER'S DISEASE

ACIDOSIS See DIABETES

ACID RAIN

A perfect proof that 'no man is an island, entire of itself' − no man, and no product of the human race, no engine, no fuel-burning power plant, no factory whose waste products are emitted into the atmosphere.

A power plant burning high-sulphur coal, belches tons of sulphur dioxide and other gases daily from its chimneys into the wind, blowing far and wide. The sulphur dioxide gas combines with water vapour in a series of chemical reactions that result in the production of sulphuric acid. This strong acid may end up in rainwater falling on a field hundreds or thousands of miles away where it upsets the chemical balance of the soil or, flowing farther, may turn a lake into a watery desert, empty of plant or animal life.

Similarly, a steady stream of traffic clogs the air with pollutants – chiefly carbon monoxide, lead, and oxides of nitrogen. The nitrogen oxides combine with water vapour in the atmosphere to form nitric acid, which falls as the not-so-blessed rain. Countries such as the United States, Japan, and Sweden now equip their cars with unleaded petrol and catalytic converters (see CATALYSIS AND CATALYSTS) in an effort to clean up this aspect of air pollution.

In the late 1980s an EEC directive approved by Britain required countries to reduce emissions of sulphur dioxide from electric generating plants by the year 2003. The US Congress was considering tightening controls over emissions from power stations and other industrial sources.

Even before people came on the scene, pollutants were present in the atmosphere, emitted by volcanoes, geysers, marshes, and soil bacteria, but in nowhere near the quantities that today produce acid rain.

ACOUSTICS

There's more to this than meets the ear, for acoustics (Greek *akouein,* to hear) includes not only the reception of sounds but also their production, transmission, and uses. *Musical acoustics* is concerned mainly with the design and production of musical instruments – and all hail Antonius Stradivarius (1644–1737) of Cremona, Italy. *Engineering acoustics* deals with the more technical aspects of sound production: sound systems and their components – microphones, headphones, amplifiers, loudspeakers. *Architectural acoustics* is the nightmare field dealing with the design of enclosed spaces, particularly auditoriums and concert halls, for the achievement of optimum sound transmission and reception – nightmare because its principles are not entirely clear, and a mistake is no easy matter to rectify. *Environmental acoustics* deals with the problems of noise control and noise pollution, an increasingly important subject in this era of growing traffic din and higher-powered electronically produced sound. *Noise insulation engineering* is a subdivision of this.

ACQUIRED IMMUNE DEFICIENCY SYNDROME   See AIDS

ACTIVE SONAR   See SONAR

ACUPUNCTURE

Originally an ancient Chinese practice for relieving pain and treating some kinds of disease. In the early 1970s acupuncture gained supporters in the Western world as a method of inducing anaesthesia without the use of chemicals. A patient under acupuncture anaesthesia remains conscious during surgery, apparently without significant pain.

In acupuncture (Latin *acus,* needle + English *puncture*) thin needles are inserted just under the skin at some of the hundreds of designated acupuncture points. These are places where networks of nerves are accessible, as in the toes, palms of the hands, earlobes, and face. The points are specific for the condition; for example, a trained acupuncturist treating facial pain places the needles at certain points on the hands and feet as well as the face. The needles are usually twirled by hand, but sometimes a low-voltage electric current is passed through them (clearly not a part of the ancient method). In one high-technology needleless variation, a LASER beam penetrates the skin to the depth usually reached by the needles.

One concern about acupuncture is that although the method is not itself harmful, it may, by relieving a patient's pain, divert attention from some serious disorder. The reasons for the pain-deadening effects of acupuncture are not known.

ACYCLOVIR *See* HERPESVIRUS DISEASES

ADENINE *See* DNA AND RNA

ADRENAL GLAND *See* HORMONE

ADRENALINE *See* ANGINA PECTORIS; HORMONE

ADRENOCORTICAL HORMONES *See* STEROIDS

ADSORPTION RATE *See* ANALYSIS

ADVENTITIA (OF ARTERIES) *See* ARTERIOSCLEROSIS

AERODYNAMICS

The science dealing with the flow of air and other gases and the motion of objects through them. The most frequent application of aerodynamic principles is in the design of aeroplanes: self-

AERODYNAMICS

propelled, fixed-wing, heavier-than-air flying vehicles. This definition includes just about every flying machine you're likely to see while your captain awaits his turn on the runway. Aerodynamics also concerns nonaeroplane flying vehicles, such as helicopters (rotating wings), gliders (not self-propelled), and balloons (lighter than air).

The exterior of an aeroplane is designed to slide through the air with the greatest of ease. This is accomplished by avoiding assaults on the surrounding atmosphere. A smooth, rounded, tapered shape insinuates the aeroplane's way with a minimum of disturbance, whereas angular, abrupt shapes would break up the air into ragged little turbulent eddies, producing friction, called *drag,* that slows up flight and wastes power. A diagram of a plane shows the streamlined shape of the body — the *fuselage* (Old French *fusel,* spindle), and of the engine covers, the *cowlings* (Old English *cule,* hood). Notice, however, that the wheels and undercarriage — the *landing gear* — are not streamlined. Streamlining is unnecessary because the landing gear is tucked out of the way of the airstream soon after lift-off.

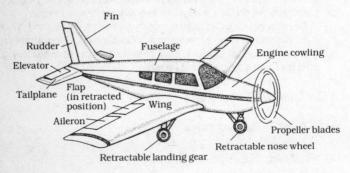

The wing is made of an ALLOY (a mixture of metals) that is stronger and more durable than aluminum and only slightly heavier.

Notice that the wing's upper surface is curved. This contributes to its AEROFOIL shape, which produces lifting power, or *lift,* when moved through the air (or when air moves against it). How this is accomplished is the subject of the following nonmathematical, too-brief points 1–4, which may be skipped and nevertheless permit on-time departure:

1. Air is a gas. Its molecules are in constant wild agitation in all directions, colliding with each other and with nearby surfaces. This motion produces air pressure.
2. Air pressure pushes down against the peaceful-looking upper wing surface. Underneath the wing, an equal air pressure is pushing upwards. The two pressures, downward and upward,

cancel out. Therefore there is no lift in the wing of a stationary aeroplane.
3. If we could reduce the downward pressure on top, the wing would begin to have lift. Reduce the pressure enough and we could overcome the pull of gravity and fly! But how to reduce the downward pressure?
4. We cleverly produce something called the BERNOULLI EFFECT. The 'we' is not editorial, for it is about to include you. Cut a 12.5- by 20-centimetre (5- by 8-inch) piece of newspaper. Hold it by two corners of one narrow side. Notice how the sheet droops in a curve, like the front portion, called the leading edge, of the wing. Now blow horizontally across the front. Watch the paper lift up vertically. A horizontal force such as your breath, exerted against a properly shaped surface, can produce a vertical effect — lift. You moved air against a stationary sheet of paper. You can get the same effect by moving the paper against stationary air, which is a closer parallel to aeroplane flight.

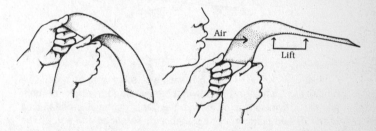

Look at the rear portion, called the trailing edge, of the wing to find two sets of movable surfaces — one set on the right wing, another on the left wing. The inboard pair are the *flaps*. These surfaces are auxiliary wings, to add to the total area of the lifting surface. The added wing area gives added lift, needed during the comparatively slow speed of takeoff. As soon as good flying speed is reached, the flaps are retracted into the hollow space in the main wing. The flaps reemerge to give added lift and braking force during the landing approach.

The outboard pair of movable surfaces are *ailerons* (French, little wings). These move in opposed fashion: When the left aileron moves up, the right one moves down, and vice versa. The effect is to *bank*, or tilt, the aeroplane during turns to prevent skidding, just as a bicycle rider leans when turning.

At the rear of the plane is the tail assembly, or *empennage* (French, feathers on an arrow). The empennage consists of two sets of surfaces, one set fixed and one set movable. The fixed set acts like

AERODYNAMICS

the feathers of an arrow, to give stability – straight flight. It consists of the *fin* and the *tailplane,* each acting in its own axis. The movable surfaces are the *rudder,* which, like the rudder on a ship, steers right and left, and the *elevator,* which tilts the aeroplane upward or downward.

TAIL ASSEMBLY

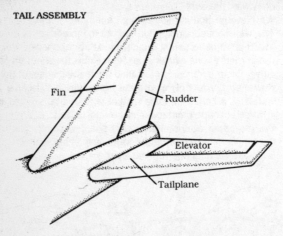

Finally, if we're to take off, we need some sort of propulsive mechanism, which fortunately is there, in the form of one or more engines, driving one or more fans. Yes, fans. Put an electric fan on a little wheeled platform like a skateboard. Turn it on and watch. The fan propels a stream of air in one direction and propels itself and its platform in the other. If you're flying on a propeller-type plane, the

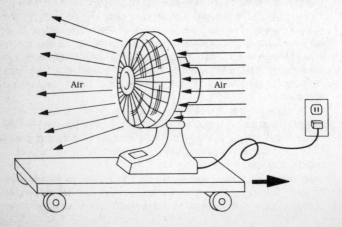

fans (one or more) are quite evident, with their big blades. Driven by engines, they push air backward and simultaneously force the aeroplane forwards. (Newton's third law of motion: a force exerted in one direction produces an equal force in the opposite direction.)

But where are the fans (propellers) on a jet plane? They're inside, hidden by the cowlings. There are several in each engine. The fans are lined up one behind the other, all mounted on the same shaft. The front fan receives outside air and forces it back to the second fan, which adds pressure to it, then to fans three and four, and so on. When the stream of air finally exits from the last fan, it is under tremendous pressure, which is boosted still further by burning the air with the engine fuel. Under tremendous pressure, the burning gases are forced past turbine blades, causing them to turn the shaft. Finally the air is blasted out to the rear in an ultrapowerful stream, or jet. Action and reaction: the jet of hot exhaust gas and compressed air exerts its force backward and simultaneously pushes the engines (and the attached plane) forwards.

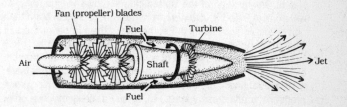

AEROFOIL

Narrowly defined, a body, such as an aeroplane wing, shaped so that its movement through air produces a lifting force; more broadly, any body shaped so that its movement through a FLUID (usually air or water) produces a desired motion or effect. Some examples of aerofoils are propellers, rudders, wing flaps, windmill vanes, and sails on a sailing boat.

See also AERODYNAMICS.

AEROPLANE   *See* AERODYNAMICS

AGC   *See* AUTOMATIC GAIN CONTROL

AGENT ORANGE   *See* DIOXINS

AI   *See* ARTIFICIAL INTELLIGENCE

AIDS

Acronym of Acquired Immune Deficiency Syndrome, a virus-caused disorder of the IMMUNE SYSTEM. The virus destroys the body's ability to fight disease-causing agents, leaving the victim susceptible to opportunistic diseases, diseases that rarely affect people with normally functioning immune systems. Examples of such diseases are *Pneumocystis* pneumonia, the leading cause of death among AIDS victims; an unusual type of tuberculosis; and Kaposi's sarcoma, ordinarily a rarely seen type of cancer. The virus may also invade and damage the brain. So far as is known, AIDS is always fatal. The AIDS virus also causes *AIDS-related complex* (ARC), a condition with less severe symptoms that often precedes full-blown AIDS.

AIDS was first identified in the United States in 1981. By mid 1988:

- 111,000 cases of full-scale AIDS had been reported to the World Health Organization of the United Nations, by 140 countries, of which more than 52,000 were from the United States. More than half of the victims had died. Estimates put the number of cases in African countries up to ten times higher than those reported.
- Hopes for early development of a vaccine had faded for a number of reasons. First, the virus changes its genetic nature (mutates) with extraordinary rapidity. Also, although the immune system produces antibodies against AIDS, as it does against other diseases, the antibodies do not protect the victim. It was expected that the next few years would see the start in Britain of trials on humans of possible components of a vaccine.
- Researchers used *azidothymidine* (AZT), a synthetic drug, to decrease the intensity of the disease and prolong life. However, they stressed that AZT is not a cure, and that it can have severe side effects. They were also testing AZT in people infected with the virus but not yet showing signs of illness. Like many other recently developed synthetic drugs, AZT is extremely expensive.
- After a slow start, trials on humans of more than 40 other drugs were under way. Some of these were designed to attack the AIDS virus, and others were intended to stimulate and strengthen the immune system. Clinical trials of a drug called *ditiocarb* showed promising results in delaying the onset of AIDS in people infected with the virus. Unlike AZT, ditiocarb appears to have only minor side effects.
- No AIDS victim was known to have regained his or her immunity naturally.
- At first, about 70% of the cases reported were young American

homosexual and bisexual men. Intravenous drug users sharing contaminated needles and haemophiliacs who received blood transfusions were also high-risk groups. But the picture changed rapidly. The rate among homosexuals fell sharply as many of them modified their sexual behaviour, while the rate among drug users rose. Screening of blood donors was introduced. In Europe and the United States the rate of infection in the general population was very low, but was rising slowly. In Africa the rate among heterosexuals was spreading rapidly.
- The virus had been identified in laboratories in the United States and in France. It has been known by several names, but the name most widely used is *human immunodeficiency virus* (HIV).
- Tests for AIDS were in use, albeit with serious limitations. The tests detect antibodies developed in response to the virus rather than the virus itself. Most individuals testing positive (HIV+) have no symptoms and may not develop AIDS for years, if at all, yet can infect others. A test may sometimes give a false positive result even when no antibodies are present. Conversely, a recently infected individual who has not had time to develop antibodies would test negative. Nevertheless, the tests are valuable, especially in screening blood intended for transfusions.

The evidence indicates that AIDS is transmitted only in semen, blood, and vaginal secretions during sexual contact, through sharing of contaminated needles among intravenous drug users, by the transfusion of contaminated blood, or by an infected mother to her baby before or during birth. Nevertheless, the very long incubation period — 5, 10, or even more years may pass after infection, before symptoms appear — the difficulty of compiling accurate statistics, and the limitations of the tests have combined to produce widespread (though entirely unrealistic) fears of infection through casual contact.

The scientific and medical problems of AIDS are paralleled by equally difficult social, political, and economic questions. For example:

- Mandatory testing for AIDS is advocated by many well-intentioned people. But who should be tested? Applicants for jobs, insurance, or marriage licences? Homosexuals? Prisoners? Prostitutes? Drug addicts? Everybody?
- In view of the relatively large numbers of false positive and false negative results, how should the test results be used? Should identities of those testing positive be revealed, possibly subjecting them to discrimination in finding housing, employment, or insurance?

- Should doctors be required to report positive results, to protect a patient's sex contacts?
- What can or should be done about people with AIDS who knowingly infect others?
- How should we deal with the problem of dentists, doctors, and other health personnel who refuse to treat AIDS patients?
- Education about 'safe sex' conflicts with the strongly held religious and moral views of many people. Can those views be accommodated in the face of an epidemic?
- Many experts advocate distribution of clean needles at no cost to drug users, to prevent infection through shared needles. Would, or do, legally elected and appointed government authorities doing this seem to sanction drug use, an illegal act? By late 1988 local health authorities in England and Wales were being encouraged and given extra finance by the government to set up schemes to provide sterile syringes.
- The cost of AIDS research, education, and treatment is enormous and increasing — by 1991 the British Medical Research Council expects to be spending over £14 million per year on research, and New York City expects to be spending $1 thousand million per year on treatment of AIDS patients. Who should pay?

The list of questions could go on and on, and unfortunately there are no easy answers to any of them.

AILERONS

Movable control surfaces on an aeroplane or glider wing, used to bank (tilt) the aircraft when making a turn.
See also AERODYNAMICS.

AIR MASS ANALYSIS

The study of the earth's atmosphere by sections (air masses) having distinct characteristics of temperature, pressure, and humidity. While such study may at first evoke a 'So what?' response, it actually represents a turning point in the understanding and forecasting of weather, known as the science of *meteorology*. Formerly, meteorology was based on surface phenomena, a kind of sophisticated 'red sky at morning, sailor take warning' approach. With the development of the air mass theory — by several meteorologists, with major credit to Norwegian physicist Vilhelm Bjerknes (1862–1951) and his son Jakob (1897–1975) — the science was placed on a three-dimensional global basis.

According to the air mass theory, there are about 20 'breeding grounds' of atmospheric conditions throughout the world. These are areas where large masses of air (800 kilometres/500 miles or more in diameter and 3.2 kilometres/2 miles or more high) remain stationary for weeks at a time, acquiring the characteristics — temperature, moisture, and density — of that area, and then move on, propelled by prevailing winds, the earth's rotation (see CORIOLIS EFFECT), buildups of pressure and temperature, and other natural forces.

One of these breeding grounds lies over the Caribbean tropical seas; its offspring emerge moist and warm — and in autumn, in their most squalling form, as hurricanes. Another source of air masses, the subarctic Canadian Plains, gestates cold, dry air masses. And there are many others, in various parts of the earth, in various seasons, giving rise to air masses of these classifications: (1) *tropical maritime* (e.g., the Caribbean); (2) *polar continental* (e.g., the subarctic Canadian Plains); (3) *tropical continental* (e.g., over the Sahara), hot and dry; and (4) *polar maritime* (e.g., over the Arctic Ocean), cold and somewhat moist. There are lesser subdivisions as well; all play roles of varying significance in the earth's major climate patterns and events.

As air masses leave their breeding grounds, they bring their meteorological conditions to the earth surface over which they travel, interacting with it, blending with local air, and gradually losing their characteristics. The weather they bring, as simple air masses, tends to be steady and uniform for several days at least; it does comparatively little to distract the forecasters from their crossword puzzles. What captures their interest most is a confrontation between two air masses, along a battle line called, naturally, a *front*.

There are four basic types of fronts: cold, warm, stationary, and occluded. The labelling is based on which air mass is claiming victory.

Cold front An advancing cold air mass forcing a warm air mass to retreat. Cold air, being denser and heavier, presses in under the warm mass in a wedge. (The wedge is much flatter than the one shown here.) Along the front, cold air causes moisture in the warm mass to condense and form clouds and perhaps rain or snow. The advance of the wedge shows up as higher and higher clouds, until finally cool, clear settled weather prevails, usually until it is succeeded by a warm front.

Warm front The forerunner of a warm air mass pressing forward against a retreating cold mass. Warm air being lighter, the front is a backward-facing wedge, squeezing its way over the cold mass. If the warm mass is moist, it is chilled along the front, forming clouds and rain. Clues to a warm front are lower and lower clouds, often

COLD FRONT

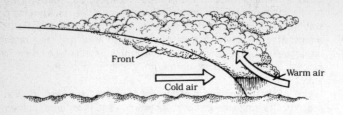

WARM FRONT

OCCLUDED FRONT

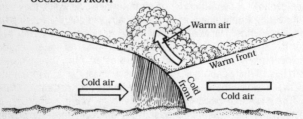

persisting until a general mean drizzle prevails. When relief arrives, it is usually brought by a new cold air mass, heralded by cold-frontal conditions. Either way, the passage of a front always brings a weather change. However, sometimes there may not be enough pressure difference between the huff of the cold mass and the puff of the warm mass to produce a satisfactory advance or retreat, in which case we have a stationary front.

Stationary front A self-explanatory, self-annihilating situation, terminating in a mingling, or homogenization, of former enemies. However, a different result may occur, a kind of *ménage à trois* called an occluded front.

Occluded front A warm mass, pushed backwards by a cold mass, and encountering another cold mass that refuses to budge. Having nowhere to go but up, the warm mass does that, supported on the strong shoulders of the two cold masses, closed off from life below (Latin *occludere,* to close off). Such a front is a stagnant bore, especially over air-polluted cities, until relief arrives.

Air masses and fronts carry identification badges — clouds of specific shapes, degrees of whiteness, altitude, and duration. A study of cloud clues is beyond the province of this book but is available in many popular-level books on meteorology.
See also WATER CYCLE.

AIRPORT METAL DETECTORS
See MAGNETIC RESONANCE IMAGING

AIR SACS   *See* EMPHYSEMA AND BRONCHITIS

AIRSHIP   *See* BALLOON

ALANINE   *See* PROTEINS AND LIFE

ALGEBRA   *See* BOOLEAN ALGEBRA

ALGOL   *See* ALGORITHM; COMPUTER LANGUAGE

ALGORITHM

A procedure involving a series of steps, often of a yes-or-no form, used in mathematics, logic, and computer programming. The term is derived from al-Khuwārizmi, the surname of the 9th-century Arab mathematician who developed the formal rules of algebra. Here are two arithmetic examples, using the algorithm for division:

$$\frac{132}{3\overline{)402}} \quad \frac{93}{3\overline{)279}}$$

> *Question:* Does 3 go into the first digit at least once? *If yes,* write the number of times, and apply the remainder, if any, to the next digit to the right. *If no,* try the first two digits, write down the result, and apply the remainder to the next digit to the right, and so on.

Yes-and-no questions, followed by command sequences, make up the algorithms in all arithmetic processes.

A meteorologist working out the day's weather forecast uses dozens of different algorithms for calculating air temperature, dew point, humidity, and so on, but his labours are much lightened by loading his computer with ALGOL, the *algorithmic-oriented* *l*anguage. ALGOL contains thousands of scientific and technical algorithms, available at a typed command.

ALLERGY   See IMMUNE SYSTEM

ALLOY

A mixture of two or more metals, to obtain or increase some desirable quality. Copper and zinc, for example, are alloyed to make brass, which is easier to machine and more resistant to corrosion. Some alloys are made by adding a nonmetallic substance to a metal. For example, iron, a brittle metal, is hardened by the addition of a very small percentage of carbon, forming steel. Another alloy, Duralumin, is widely used in aircraft because it is stronger, more durable, and only slightly heavier than aluminum. It is mainly aluminum, with small amounts of copper, magnesium, and manganese.

ALPHA CELL   See DIABETES

ALPHANUMERIC

Containing both alphabetical and numerical information. Your telephone bill is alphanumeric. A love letter is usually not.

ALPHA PARTICLE

The nucleus of the helium atom, consisting of two protons and two neutrons. Several substances undergoing radioactive decay emit these particles.
   See *also* ELEMENTARY PARTICLES; RADIOACTIVITY.

ALPHA WAVES   See ELECTROENCEPHALOGRAPH

ALTERNATING CURRENT   See DC AND AC; RADIO

ALTERNATING CURRENT, GENERATION OF
See GENERATORS AND MOTORS, ELECTRIC

ALTERNATOR

Any electrical generator that produces an alternating current, or AC. Let us look at one special kind of alternator, the type used in motor cars.

Formerly, a car battery was used only to provide electric power for lights, the ignition system, starting, and similar basic jobs. The battery had to be recharged. This could be done only with direct current (DC), so a DC generator, driven by the car's engine, was used.

Today, the basic jobs, plus electric windows, brighter headlights, cassette players, rear-window defrosters, fans, and other power-hungry conveniences, demand much more electricity. A DC generator's output is not adequate during stop-and-go and idling speeds. An AC generator could do the job, if only it produced the right kind of current. Enter the alternator. It generates AC, which is changed to DC by a rectifier, a device that allows current to flow in one direction only.

See also DC AND AC; GENERATORS AND MOTORS, ELECTRIC; SEMICONDUCTOR.

ALVEOLI   See EMPHYSEMA AND BRONCHITIS

ALZHEIMER'S DISEASE

A brain disorder occurring mainly among elderly people; it has also been called, or associated with, senile psychosis, senile dementia, and senility.

One prominent symptom of the disease is progressive loss of memory (this, however, can also be a normal part of aging). The patient suffers from impaired judgment, confusion, and irrational ideas.

The cause of the disease is unknown. One proposed explanation which has received a lot of attention holds that Alzheimer's is an autoimmune disorder, in which the body's IMMUNE SYSTEM causes the damage.

Certain kinds of cells in the brain of a person with Alzheimer's disease atrophy (waste away) and die. These cells are known to produce a neurotransmitter called *acetylcholine*. Neurotransmitters are substances that are involved in generating and transmitting nerve impulses. Without them, nerve pathways cease to function.

Until recently the course of Alzheimer's disease was thought to be inevitably downhill; now certain drugs seem to promise some help. For example, it appears that *naloxone,* used in treating overdoses of narcotics, may be of use in treating loss of memory. Another drug, *tetrahydroaminoacridine,* shows great promise; it acts to slow down the rate at which acetylcholine is broken down. Clinical trials of the drug in America were suspended after it appeared to cause liver damage in some patients, but resumed in 1988 using lower dosages.

AM   See RADIO

AMINO ACIDS See PROTEINS AND LIFE

AMNIOCENTESIS

A technique for removing some of the fluid that surrounds a fetus in its mother's uterus (womb). At first it was used as a way of relieving pressure caused by an excess of fluid, but today it has become a valuable tool in *genetic screening* — the diagnosing of many kinds of disorders in a fetus, long before birth. It also makes it possible to learn the sex of the fetus.

An *embryo* (it is promoted to the rank of *fetus* after the second month) begins as a fertilized egg cell. The cell attaches itself to the inner wall of the uterus, a hollow muscular organ in the mother. Around the embryo grows a thin, membranous sac — the *amnion*. The embryo lives in the amniotic cavity that is formed, bathed in amniotic fluid.

In performing amniocentesis (Greek *amnion* + *kentein,* to prick), a hollow needle is carefully inserted into the amniotic cavity and some fluid is withdrawn. The fluid holds some of the cells that an embryo normally sheds as it develops. The cells are separated from the fluid and grown by TISSUE CULTURE to provide a mass of cells sufficient for analysis. Microscopic examination of a cell's chromosomes can reveal the sex of the fetus and the presence of some types of disorders, notably DOWN'S SYNDROME. Chemical analysis of the cell culture and amniotic fluid can spot the presence of certain GENETIC DISEASES.

Amniocentesis is a specialized variation of a technique called *paracentesis,* which is used to remove unwanted fluid from body cavities where it may accumulate — for example, in the abdominal cavity or in the pericardial cavity around the heart.

Chorionic villus sampling (CVS) is a newer way to obtain fetal cells. It provides larger samples, can be done earlier in pregnancy, and requires no needle. The sample is taken directly through the uterus, where the chorionic villi — tiny fingerlike projections arising from the outer layer of the fetus — provide a bridge for the exchange of materials between fetus and mother.

See also DNA AND RNA.

AMNION See AMNIOCENTESIS

AMPERE See ELECTRICAL UNITS

AMPLIFIER

A device for increasing the strength (voltage) of electrical signals. In radio receivers and television sets, the signals are very feeble electric currents (a few millionths of a volt), produced by radio waves sweeping across an aerial. Amplifiers, powered by batteries or house current, boost these signals to a sufficiently high voltage to operate a loudspeaker, from 0.5 volt (a small portable radio) to 30 volts (a rock concert speaker), or to form a picture on a television tube, which requires 20,000 volts or more. Amplification was once done by vacuum tubes, but these have been replaced by TRANSFORMERS and transistors.

See also SEMICONDUCTORS

AMPLITUDE MODULATION   See RADIO

AMYOTROPHIC LATERAL SCLEROSIS
See NEUROMUSCULAR DISORDERS

ANABOLIC STEROIDS   See STEROID

ANALOGUE   See DIGITAL AND ANALOGUE

ANALYSIS

The identification of a complex substance or form of energy. This is usually done by separating it into its various components.

The world consists mostly of mixtures. This paper is a mixture of fibres, binders, and bleaches. The light falling on it is a mixture of coloured beams that collectively approximate the sensation of whiteness. The air you breathe is a mixture of thousands of different gases, vapours, and solid particles. In fact, in this mixed-up world, the concept of nonmixedness, of 'purity', is almost an abstraction. Yet there are occasions, especially for scientists, when a substance or form of energy must be spread out, separated into its components in order to identify them, to extract a desirable ingredient, or to reject an undesired one. Towards these purposes, scientists have developed numerous separation and identification techniques that are collectively categorized as analysis (Latin, to loosen or undo). Here are examples of some common types:

Filtration An analytic technique that makes use of differences in particle size. Filter paper separates coffee grounds from liquid coffee, red blood corpuscles from blood fluid (plasma), and many other such mixtures of solids and liquids. Unglazed porcelain is another filtration medium used in the same way.

Centrifugation A technique that makes use of differences in density (heaviness). In its most primitive form it enables Klondike

ANALYSIS

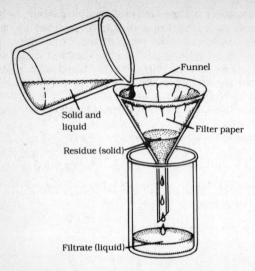

Joe to scoop up a handful of streambed sand, drop it into a panful of water, swirl it vigorously, and hold his breath, eagerly waiting for grains of gold to settle to the bottom first, gold being about seven times as dense as sand. The same principle is the basis of many laboratory devices, the most common being the *centrifuge,* in which the mixture to be analysed is placed in a small container and spun at great speed, as much as 4000 revolutions per second. The spin produces a force, called centrifugal force, thousands of times greater than gravity. This exaggerates the differences in density and therefore separates particles that are even minutely different. Mixtures of liquids, such as cream and milk, can also be separated by centrifugation.

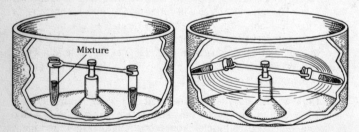

Chromatography A technique that takes advantage of a difference called the *adsorption rate,* which might be loosely defined as degree of stickiness, the tendency of one substance to cling to another. Here's a simple demonstration of the principle. First, two colours of ink are mixed, and a drop of the mixture is placed on a strip of uncoated paper more than 25 millimetres (about ½ inch) from one end. Then the strip of paper is set upright in 25 millimetres (½ inch) of water. As the water soaks up into the dry paper, it passes through the ink mixture. The different ingredients of the mixture have different clinging tendencies for the paper — different adsorption rates — so they become selectively removed or unmixed as they are dissolved by the rising water. The result is a series of coloured bands, one for each of the ingredients of the mixture. Because this process was first used for analysing mixtures of different-coloured ingredients, it was named chromatography (Greek *chroma,* colour), but it is also used for analysing mixtures of same-coloured ingredients.

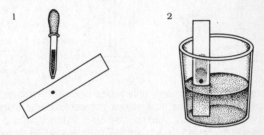

Electrophoresis A related process (Greek *phoresis,* being carried) in which differential adsorption is also involved. While the water is being absorbed vertically an electric current flows sideways across the paper. Because the various ingredients respond differently to electricity, they are separated more rapidly and more precisely than by water alone.

Distillation A technique based on differences in boiling point that is used in separating mixtures of liquids. A mixture is heated just barely enough to reach the boiling point of the easiest-to-boil (most volatile) liquid. As it boils away, its vapour is caught and cooled, to condense into a pure liquid all by itself. Then the remaining mixture is heated to the next highest boiling point, the process is repeated, and so on. The distillation apparatus is called a still. The largest use of distillation is not, as you might suppose, in making whisky, which is large enough, but in refining petroleum (Greek *petra,* rock + Latin *oleum,* oil), separating it into petrol, paraffin, lubricating oils, heating fuels, diesel oil, and many other substances. What material remains is black, sticky, and useful for paving — asphalt.

Chemical analysis Testing to identify substances. Klondike Joe's hopes may have been raised in vain by the deceptive yellow glitter of

ANALYSIS

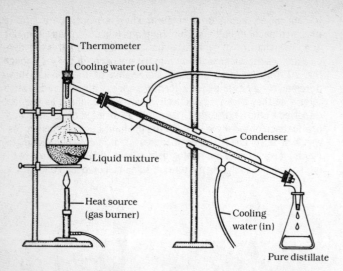

fool's gold — iron sulphide — only to be dashed at the assay office by the application of a chemical reagent, hydrochloric acid. Pure gold is unaffected, but fool's gold seethes and bubbles in a chemical reaction that converts it into a liquid, iron chloride solution. There are thousands of chemical tests for analysis of elements, compounds, and mixtures.

Chemical analyses are of two types: *qualitative* and *quantitative*. Qualitative analysis (Latin *qualis*, of what kind) tells of the presence or absence of a substance, as in the fool's gold test. Quantitative (Latin *quantis*, how great) analysis tells us how much of each ingredient is present. Probably the most ubiquitous and unglamorous example of chemical analysis is printed on the side of a cereal box.

The analytic techniques described thus far — filtration, centrifugation, chromatography, electrophoresis, distillation, and chemical analysis (there are many more) — all have to do with the analysis of substances. Forms of energy, too, can be analysed: separated into their components and examined individually. Let's begin with an energy analysis that you have probably performed once or twice today:

Frequency analysis The space around you swarms with hundreds of RADIO programmes coming from broadcasting stations, police transmitters, radio hams, walkie-talkies, and other disturbers of the ethereal peace, all clamouring for admission into your receiver. These programmes come riding in on radio waves that differ in

TYPICAL NUTRITIONAL COMPOSITION PER 100 GRAMMES	
Energy	302 kcal 1,285 kJ
Protein	11.4g
Fat	4.2g
Dietary Fibre	13.6g
Available Carbohydrate	66.9g
Vitamin C Niacin Vitamin B_6 Riboflavin (B_2) Thiamin (B_1) Folic Acid Vitamin D Vitamin B_{12}	35.0mg 16.0mg 1.8mg 1.4mg 1.0mg 250 µg 2.6 µg 1.7 µg
Iron	38.0mg

frequency (the number of waves per second). Each station has its own assigned frequency. The tuning knob, dial, and circuitry of your receiver have the job of selecting from that jumble only those waves whose frequency you have chosen. Turn the knob a few degrees and you pick up the next frequency, and the next, and so on. In effect, a radio tuner is an analyser of radio wave frequencies.

In Arecibo, Puerto Rico, there is a highly sensitive radio telescope (i.e., a specialized radio receiver) connected to a dish-shaped antenna 7.284 hectares (18 acres) in area. It analyses the millions of different radio signals being received constantly from the solar system and outer space. These signals are not music or drama programmes, not even soap commercials (although such would be received with mixed astonishment, delight, and dismay), but a cacophony of peeps, squawks, rumbles, and sputters, emitted by the ingredients of the busy universe — glowing particles in the sun's corona, carbon dioxide on the polar caps of Mars, vaporized iron in one of the stars of Ursa Major; all these and millions more emit their radio frequency signatures, which are picked up and analysed in Arecibo and elsewhere on earth by radio analysers that are more expensive and sensitive, but not too different in principle, from the little box that wakes you with the morning news.

Nuclear magnetic resonance (NMR) From the enormous Arecibo receiver, to your little bedside radio, we go now to one of the tiniest radio receivers in the universe: the nucleus of a single atom. The nucleus of each kind of atom has its own 'natural' frequency; it is tuned in, so to speak, to one station only. Send in the particular frequency of hydrogen, and hydrogen nuclei respond by resonating (see RESONANCE). We can read on a meter, or see on an oscilloscope, that it has been accepted. Send in a different frequency (of oxygen, for example), and hydrogen nuclei will not accept it. Because every kind of atomic nucleus responds (resonates) to its own unique frequency, these frequencies are signatures of the atomic elements in a mixture. With the added use of a strong magnetic field, substances can be analysed for their composition, a process called nuclear magnetic resonance.

Here, for example, are the signatures on an oscilloscope screen of the hydrogen atoms in ethanol, the foundation of every alcoholic drink. Ethanol, also known as ethyl alcohol, or just alcohol, is composed of carbon, hydrogen, and oxygen atoms attached in this way: $CH_3 - CH_2 - OH$.

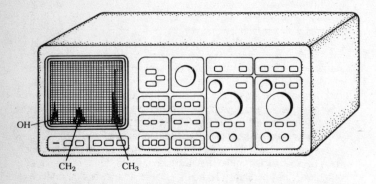

NMR has the great advantage of being nondestructive, so it can be used on living things without harming them. Using this method of analysis, a scientist can even follow the progress of a tiny amount of a particular substance as it passes through a series of chemical changes within a living organism (see MAGNETIC RESONANCE IMAGING).

Spectroscopy This method of analysis could be referred to as the dissection of rainbows. Hold a glass prism, or any transparent, sharp-cornered object (a 20-carat diamond will do nicely), in sunlight and observe the rainbow it casts on the nearest surface. Try it in the beam of a car's headlight (the light source is a glowing

ANALYSIS

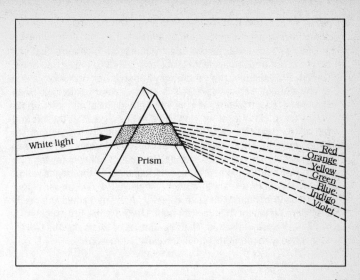

tungsten filament) and you get a different rainbow — different by the presence or absence of various colours and in the proportion of each. Every light source produces its own particular rainbow, or *spectrum* (Latin, image), depending on which glowing substances are emitting light. Seen through a *spectroscope,* the broad, soft-edged rainbow

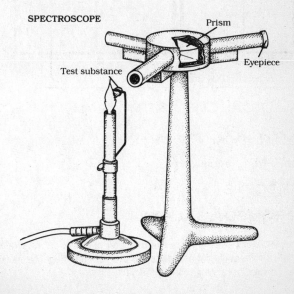

ANALYTICAL CHEMISTRY

bands of coloured light become dissected and focused into narrow, sharp lines of precise colour and location.

Every substance has its own spectrum, its own 'signature' that can be identified through a spectroscope or graphed on a spectrographic screen. For example, the yellow light from burning driftwood is due to the element sodium (sea salt is mostly sodium chloride). Seen through a spectroscope, the yellow light divides into two bright yellow lines, close together, and two dim lines, green at the left and red at the right. No other substance has exactly that spectrum. In fact, when in 1868 the British astronomer Joseph Lockyer was studying the sun's gaseous envelope (corona), finding many familiar spectra of earthly elements, he also found, near the bright yellow pair of sodium lines, another bright single yellow line, never before reported. In honour of the sun (Greek *helios*), he named the newly discovered element helium. Not until 1895 did the British chemist William Ramsay discover that helium also exists in the earth, associated with uranium, and in the earth's atmosphere.

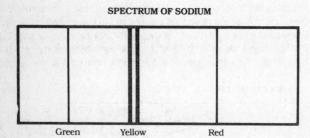

SPECTRUM OF SODIUM

Green　　Yellow　　Red

ANALYTICAL CHEMISTRY　See CHEMISTRY

ANAPHYLAXIS　See IMMUNE SYSTEM

ANATOMY　See BIOLOGY

ANDROGENS　See STEROIDS

ANDROMEDA GALAXY

A spiral GALAXY nearly twice the diameter of our Milky Way galaxy, which it resembles. At a distance of over 2 million light-years, it is the second closest neighbour of the Milky Way.

ANEURYSM

A permanent widening of an artery whose walls have been weakened by infection, injury, birth defect, ARTERIOSCLEROSIS, or HYPERTENSION. The term comes from Greek *aneurynein*, 'to dilate'. Aneurysm of the aorta is a serious condition; the aorta is the principal artery, the conduit through which blood is pumped by the heart and then distributed to lesser arteries throughout the body. A ballooning aortic aneurysm may rupture, with a life-threatening loss of blood into the body cavities. Surgical replacement of the weakened part of the aorta with a tube of polyester gives very good results.

An aneurysm in an artery of the brain is potentially life-threatening. Surgeons can lessen the danger of bursting by the use of special kinds of clips to clamp an affected artery. Another method is to wrap plastic material, or tissue taken from another part of the body, around the artery.

Physical examination may reveal some kinds of aneurysm; more often, the diagnosis is made by X-ray, sonogram (see ULTRASONICS), or CAT scan (see COMPUTERIZED AXIAL TOMOGRAPHY).

ANGINA PECTORIS

A condition named from Latin words meaning 'strangling in the chest' in which the heart suffers a shortage of oxygen. Usually this happens because the heart, even as it pumps oxygen-laden blood to all parts of the body, receives insufficient amounts of blood from its own suppliers, the *coronary arteries*. (A case of the shoemaker's child going barefoot, if you don't push the analogy too hard; see HEART ATTACK).

The angina patient has bursts of chest pain and a feeling of suffocation. Anginal pain is most often brought on by exertion and can be quickly relieved by rest. Nitraate medications that relax the heart muscle and dilate the blood vessels can produce dramatic relief in minutes. The best known of these drugs is *nitroglycerin*. A newer group of drugs, called *beta blockers*, act against adrenaline, a naturally secreted HORMONE that stimulates the heart. The result is a slower heartbeat, reducing the heart's need for oxygen.

Beta blockers are valuable in treating other conditions of the circulatory system, such as severe disturbances of the heart rhythm (*arrhythmias*) and HYPERTENSION.

ANGSTROM

A unit of length, symbol A or Å, named after Anders Jonas Ångström, a Swedish physicist (1814–1874), who was noted for his study of light, especially spectrum ANALYSIS. He discovered hydrogen in the sun's atmosphere.

An angstrom is one hundred-millionth of a centimetre. (Typing paper is about 1 million angstroms thick.) Angstrom units are most frequently used in measuring light waves. Waves of visible light, from deep red to violet, range from about 4000 to 7000 Å per wave.

ANNULAR ECLIPSE   See ECLIPSE

ANOREXIA NERVOSA AND BULIMIA

Psychiatric disorders connected with eating. *Anorexia* is from the Greek *an,* 'without' + *orexia,* 'appetite'; *bulimia* is from Greek *boulimia,* 'great hunger'. Seemingly opposites, the two conditions have more similarities than differences and may even be observed in the same person.

Quite often victims are young, female, and from middle-class families whose members are closely involved with one another. Both anorexics and bulimics suffer from low self-esteem, are obsessed with food and diets, and are intensely afraid of becoming fat. The anorexic's pursuit of slimness goes pathologically far beyond ordinary dieting, to the point of malnutrition and even death. Psychiatrists generally agree that this bizarre behaviour is based on the anorexic's need to establish a sense of selfhood and control over her or his own life.

Similarly, the bulimic suffers from an irresistible craving for food. She or he may eat a week's supply of food in a few hours; most of it is lost by natural or self-induced vomiting or by the use of laxatives. Nevertheless, enough food is absorbed so that no weight is lost.

These conditions require psychiatric help, as well as medical attention for the dangerous physical problems they cause.

ANTIBIOTICS

Chemicals produced by living *microbes* (moulds and bacteria) and named from Greek words meaning 'against life'. Antibiotics are widely used in treating certain diseases. Nobody knows how many millions of lives have been saved by enlisting these chemicals in the fight against disease-causing (*pathogenic*) microbes. The idea isn't new. Thousands of years ago the Egyptians and Chinese and the Indians of Central America used poultices of mouldy, rotting materials to treat skin ailments and infected wounds. Knowing nothing of microbes, the ancients understandably hailed the results as magic; yet even now, antibiotics are sometimes called 'miracle drugs' because they cure so many diseases regarded as incurable only half a century ago. Plague, tuberculosis, syphilis, cholera, typhoid — these and other scourges yield to antibiotics.

The era of modern antibiotics dates from the late 1920s, when Dr Alexander Fleming, a Scottish bacteriologist, observed that a certain mould produced a gold-coloured liquid that inhibited the growth of a pathogenic bacterium *Staphylococcus aureus*. 'Staph' is a cause of boils, food poisoning, blood poisoning, and bone and kidney infections. The mould was later identified as *Penicillium notatum,* a relative of the moulds that inhabit stale bread and blue cheeses, and the liquid was named *penicillin.* Fleming's success against the staph bacterium was confined to test tubes in the laboratory, but he suggested that penicillin might be used against infections in the body. Later, after methods for producing enough penicillin were developed (although extremely small amounts are needed), physicians found that it was astonishingly effective in treating pneumonia and scarlet fever, yet it worked poorly or not at all with some other infections. In time, new antibiotics were found that filled in the gaps left by penicillin. Nevertheless, scientists pursued the goal of broad-spectrum antibiotics — single medicines that would be effective, in shotgun fashion, against a wide range of pathogens

Today, there are many broad-spectrum antibiotics, such as the cephalosporins and tetracyclines, but scientists continue to hunt for new ones, for a good reason. Like other living things, bacteria change or mutate: a new generation may show some trait not possessed by previous generations. Thus a strain of pathogenic bacteria may appear that is able to live in the presence of a particular antibiotic. While normally susceptible bacteria succumb, the mutant group, free of competition, flourishes, to the distress of the infected patient. This problem can be dealt with by switching to some other antibiotic or by prescribing a combination of antibiotics. For the long-range solution, scientists must work to develop a new kind of antibiotic that can inhibit the new strain of bacteria. Further down the road, however, still another anti-antibiotic mutation may occur, and the need for still a newer antibiotic, and so on.

Of the more than 1000 antibiotics that have been produced, about 100 have proved to be both effective and nontoxic for human use.

Laboratory observations offer clues to the way antibiotics may work in the body. Some kinds of antibiotics actually dissolve bacteria, but most operate with a less violent chemistry. Some interfere with the ability of bacteria to absorb or use nutrients; others prevent the building of cell membranes and walls.

Antibiotics are of enormous value in controlling the epidemics that once killed millions of domestic animals, but their use in the farmyard extends beyond disease. The growth of livestock is greatly speeded up by routine administration of antibiotics. However, there is considerable disagreement about the wisdom of this practice. Some experts fear that antibiotics, by killing off susceptible bacteria in livestock, may encourage an increase in resistant varieties harmful to humans. In some parts of the world (not in Britain), meat and fish are treated with antibiotics to prevent spoilage.

Nowadays, antibiotics are produced in three ways:

1. Selected microbes are provided with the food and living conditions calculated to bring out their best efforts.
2. *Synthetic antibiotics* are made entirely by chemical manipulation in the laboratory.
3. *Semisynthetic antibiotics* are the product of a remarkable partnership in which microbes make antibiotics that are then altered chemically in the laboratory to endow them with various desirable qualities.

ANTIBODIES See IMMUNE SYSTEM

ANTIGENS See IMMUNE SYSTEM

ANTIHAEMOPHILIC FACTOR See HAEMOPHILIA

ANTIHISTAMINES See HISTAMINE

ANTI-INFLAMMATORY AGENTS See STEROIDS

ANTIMATTER

Matter made of *antiparticles*, which are similar to the particles of our familiar world — the molecules and atoms and their constituents of which we and church steeples are made — but with certain important differences. An atom of ordinary hydrogen, for example, consists of a heavy, positively charged (+) proton, orbited by a lightweight, negatively charged (−) electron. An atom of antihydrogen would consist of a heavy, *negatively* charged proton, called an *antiproton*, orbited by a lightweight antielectron that is *positively* charged and is therefore called a *positron*. Such an atom of antimatter would weigh the same and apparently behave the same if it were all by itself. If a church steeple were assembled all by itself out of antimatter, you couldn't tell the difference. Until . . .

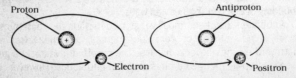

This is a favourite science fiction ploy. A male explorer from our world encounters, via TV, a female from an antimatter world, not knowing her origins. Love dawns between them, and they agree to meet somewhere in extragalactic space. They approach, embrace, and poof! they disappear in a blinding flash of light, heat, infra-red rays, cosmic rays — the works. His matter and her antimatter have cancelled one another and turned into radiant energy, their love never to be consummated.

Actually, antimatter does exist, and these annihilations do occur, though not so romantically. Antiprotons, positrons, antiquarks — a host of antiparticles — are produced in nature during the complex process of changing solar material into sunlight and radiating that sunlight. Antimatter is produced artificially during the operation of a PARTICLE ACCELERATOR (an 'atom smasher'). The antiparticles exist for a billionth of a second, or thereabouts, until they encounter a particle of ordinary matter and the two disppear in a glorious flash.

Proof that antimatter isn't just grist (antigrist?) for the science fiction mills is a technique called *positron emission tomography* (PET for short), a highly sophisticated cousin of the CAT scan. In place of the X-rays employed in CAT, PET uses the radiation that results from the meeting of positrons and ordinary electrons (the glorious flash mentioned earlier). PET, like CAT, furnishes sharp images of organs within the body, but it goes beyond that. It enables scientists to analyse the chemical processes within living inner organs, especially the brain, without harming them. It thus becomes a powerful tool in locating and diagnosing brain tumours and helps scientists in researching the brain structures and processes underlying such mental diseases as epilepsy, depression, and Parkinson's disease.

See also ELEMENTARY PARTICLES.

ANTIPARTICLE See ANTIMATTER

ANTIPROTON See ANTIMATTER

ANTIQUARK See ELEMENTARY PARTICLES

AORTA See CIRCULATORY SYSTEM

APPARENT MAGNITUDE See MAGNITUDE

AQUIFER See GROUNDWATER

ARC (AIDS-RELATED COMPLEX) See AIDS

ARCHITECTURAL ACOUSTICS See ACOUSTICS

ARRHYTHMIA See ANGINA PECTORIS

ARTERIOLE See CIRCULATORY SYSTEM

ARTERIOSCLEROSIS

A disease of the arteries. An artery is made rather like a well-made garden hose. Both are thick-walled pipes, composed of several layers that surround a central hole, or *lumen*. A liquid under pressure flows through the lumen, water in one case, blood in the other. In the case of the arteries, the pressure comes from the pumping action of the heart. An extremely thin, smooth layer of cells, called the *intima*, lines the lumen, allowing the blood to flow with a minimum of resistance. The intima is surrounded by a muscular, elastic layer, the *media*, which in turn is surrounded by the *adventitia*, a layer of muscle and connective tissue. With each surge of blood from the heart, the artery dilates, then constricts, helping to smooth out the flow.

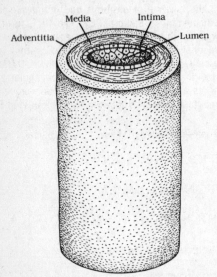

In arteriosclerosis (Latin for 'artery' + 'hardening'), the layers thicken and lose some of their elasticity, hampering the flow of blood. These changes arise from aging and from the deposition of various substances in the layers. Most important is the accumulation of *atheromas,* plaques of calcium and cholesterol. The condition is called *atherosclerosis.* Sclerotic changes are potentially dangerous because narrowed or obstructed arteries in the heart or brain may be unable to supply enough blood to those vital organs, contributing to HEART ATTACK or STROKE.

Scientists recognize that certain conditions, or *risk factors,* increase the likelihood of developing atherosclerosis. Among these factors are HYPERTENSION, a sedentary life style, cigarette SMOKING, a high concentration of CHOLESTEROL and similar substances in the blood, DIABETES, and obesity.

ARTERY   *See* CIRCULATORY SYSTEM; ARTERIOSCLEROSIS

ARTHRITIS

A disease of the joints, a leading cause of disability and crippling, though rarely life-threatening. There are many forms of arthritis (Greek *arthron,* joint + *-itis,* inflammation), but the two most common, by far, are osteoarthritis and rheumatoid arthritis.

Osteoarthritis   A wear-and-tear condition of the joints, also called *degenerative joint disease.* It is found to some extent in most people over the age of 50. In normal joints, smooth motion is made possible by a lubricant, *synovial fluid,* produced in pockets of cartilage tissue at the ends of the bones. In an arthritic joint the cartilage disintegrates; in the absence of the fluid, motion is limited and can become very painful. Weight-bearing joints such as the knees, the hips, and the spine are the most vulnerable, especially in obese people or people who have done a lot of hard physical labour.

Rheumatoid arthritis   An arthritic condition that affects up to 4% of the population, at some time, both young and old. The cause is unknown, although it may involve faulty operation of the body's immune system. The lungs, nervous system, and other parts of the body may be affected, although the joints are usually the major problem.

Swelling, redness, and pain in the affected joints are typical symptoms of arthritis. For a long time, aspirin, usually in large doses, was the only drug used to treat the pain. Now, newer drugs, used in lower dosages, are often more effective and better tolerated by many patients. Cortisone, ibuprofen, and indomethacin are among these. Rest and heat are important, and a programme of properly designed exercises can help to strengthen the joints and retain relatively good motion. Surgery is effective, and in many cases diseased joints are successfully replaced by artificial implants. No cure is known; this fact, and intense pain, sometimes drive victims to quack practitioners who offer so-called miracle cures.

ARTIFICIAL INTELLIGENCE

Sufficient creative reasoning power in a machine to perform mental tasks previously regarded as capacities only of human consciousness. This is a concept dear to many science fiction writers, the bane of theologians, and the subject of serious current research by psychologists and avant-garde computer programmers, who refer to it as simply AI.

A machine with true artificial intelligence has not yet been developed. An electronic chess player is not artificially intelligent because its programming consists merely of mathematical ALGORITHMS that test out various possible moves and evaluate their consequences, a million per second. Superior memory capacity and processing speed may someday enable this machine to defeat the human world's chess champion, but it is still not intelligent; it is simply performing routines and subroutines, standard and repetitive, based on clearly defined yes-no choices created by a human programmer.

If a machine could make choices as a result of thoughtful consideration, testing and weighing possibilities that originate within it, according to new ideas not previously conceived by its programmer, then and only then, could it be called 'intelligent'.

See also BOOLEAN ALGEBRA.

ARTIFICIAL SWEETENERS *See* SWEETENING AGENTS

ASCII

Acronym for *A*merican *S*tandard *C*ode for *I*nformation *I*nterchange. ASCII is a common code of letters, numbers, and symbols used in computers. It is a distant relative of the Morse code, which is assembled out of dots and dashes. ASCII is assembled out of binary digits: zeros and ones taken seven at a time. Capital *A* is coded as 1000001, and lowercase *a* is 1100001. The 1 at the extreme right indicates the first letter of the alphabet. The 1 at the extreme left is a code that indicates 'What follows is a letter, not a number'. The 1 next to it indicates 'This is a lowercase letter, not a capital'.

See also BINARY NUMBER SYSTEM; BIT.

ASPARTAME *See* SWEETENING AGENTS

ASTEROID See SOLAR SYSTEM

ASTEROID BELT See SOLAR SYSTEM

ASTROLOGY

A practice (its practitioners call it a science) based on the belief that the movements and relative positions of the sun, moon, and planets influence human affairs. Some astrologers also believe they can foretell the future from these celestial bodies. They maintain that given the date and time of a person's birth, they can know that person's character and personality traits and foretell his or her illnesses and length of life and the best times to marry, go into business, and make or avoid major decisions.

Ancient astrologers made valuable contributions towards the evolution of modern-day astronomy. Today, however, many astronomers would debate whether astrology fulfils the criteria of a science — for example, it lacks experiments with controls that can be duplicated by people other than the claimants.

ASTRONOMICAL UNIT See SOLAR SYSTEM

ASTROPHYSICS

The study of the origin, composition, and interactions of heavenly bodies, involving all forms of matter and energy in the universe.
See also PHYSICS.

ATHEROMA See ARTERIOSCLEROSIS

ATHEROSCLEROSIS See ARTERIOSCLEROSIS

ATOM

A unit of matter, consisting of a nucleus of positively charged *protons* and uncharged (neutral) *neutrons,* orbited by negatively charged *electrons.* A complete atom has the same number of electrons as protons, balancing the charge. A hydrogen atom, with one proton, has the lightest weight and is the only atom with no neutrons. Atoms were once regarded as indivisible.
See also ELEMENTARY PARTICLES.

ATOMIC ENERGY See NUCLEAR ENERGY

ATOMIC NUCLEUS, DISCOVERY OF
See ELEMENTARY PARTICLES

ATOMIC NUCLEUS, FISSION AND FUSION OF
See NUCLEAR ENERGY

ATOM SMASHER
See PARTICLE ACCELERATOR

ATRIA OF THE HEART
See CARDIAC PACEMAKER

AU
See SOLAR SYSTEM

AUDIO FREQUENCY
See RADIO

AURORA

A display of glowing light in the night sky, mainly in regions near the earth's magnetic north and south poles. The *aurora borealis* (northern lights) is most spectacular in Canada, Alaska and northern Scandinavia, while the *aurora australis* (southern lights) is at its best in Antarctica. The moving, shimmering glow may resemble clouds, arches, or draperies of red, yellow, and green.

Auroras are probably produced when streams of charged particles (mainly electrons and protons) from the sun collide with molecules of nitrogen and oxygen in the atmosphere, around 100 kilometres (60 or more miles) above the earth. The charged particles are trapped by the earth's magnetic field and guided towards its strongest areas, the north and south magnetic poles.

The orange glow of a neon sign is produced in much the same way, by the collision of molecules of neon gas with electrons supplied by an electric current.

See also SOLAR WIND.

AUTOIMMUNITY
See IMMUNE SYSTEM

AUTOMATIC GAIN CONTROL (AGC)

An electrical circuit in a radio or television receiver that automatically controls the level of sound coming from the loudspeaker. Also called automatic volume control (AVC), it is needed because the strength of the signals from different stations varies. The signals weaken with distance and when partially blocked by mountains and high buildings. Atmospheric changes, especially at sunrise and sunset, also alter the signals. Further, the broadcasting equipment at different stations varies in power. For some or all of these reasons, there would be a big difference in sound level from one station to another if AGC did not automatically regulate the different signal levels to produce the same, or nearly the same, sound levels.

Even after you choose a station and adjust the sound to your liking, it could change with changing atmospheric conditions. AGC keeps the sound at the chosen level. Another use of the AGC type of circuit is to hold the brightness and quality of TV pictures at an even level.

See also LOUDNESS CONTROL.

AUTOMATIC VOLUME CONTROL (AVC)
See AUTOMATIC GAIN CONTROL

AUTONOMIC FUNCTIONS *See* BIOFEEDBACK

AVC *See* AUTOMATIC GAIN CONTROL

AZIDOTHYMIDINE *See* AIDS

AZT *See* AIDS

BACKGROUND RADIATION

Radiation (radio and infrared waves) indicating that the temperature of 'empty' space is not ABSOLUTE ZERO, as could be expected, but almost 3°C (5.4°F) above that. Like the warm air near a fire that has gone out, this background radiation is regarded as evidence of energy probably remaining from the time of the beginning of the universe.

See also COSMOLOGY.

BACTERIAL VIRUSES   See VIRUSES

BACTERIOPHAGE   See VIRUSES

BALL LIGHTNING   See THUNDER AND LIGHTNING

BALLOON

An aircraft that is lighter than air and is instantly recognizable by its large gas bag. The peculiar-seeming state of being 'lighter than air' means only that the total weight of the bag, the light gas that fills it (helium, for example), the passengers and equipment, and the basket they occupy is less than the weight of the air pushed aside (displaced) by them.

The balloonist gains altitude by dropping ballast (usually loose sand or water, to avoid damage to the earthbound populace). Releasing controlled amounts of gas from the bag permits a descent. Balloons have no means of propulsion and so are at the mercy of the winds. However, *airships* (or *dirigibles*), which are cigar-shaped balloons with engine-driven propellers and steering rudders, are capable of fully directed flight.

BAR CODE   See UPC

BAROMETER   See BAROMETRIC PRESSURE

BAROMETRIC PRESSURE

One of the most common terms in weather reporting, but least understandable by commonsense reasoning. Common sense indi-

BAROMETRIC PRESSURE

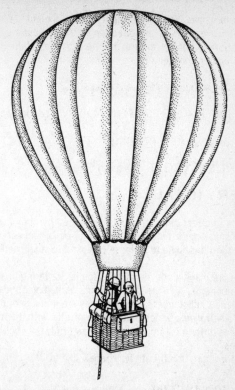

cates that a moist handkerchief is heavier than a dry one. Yet moist air is lighter than dry air. Cautiously, now . . .

Scoop up a bucketful of dry desert air and one of moist air from over the frothy waves of the ocean. Count the number of molecules in each bucket (this can be, and has been, done, but not by simple means). The same number of molecules is in each, but the proportions are different. Dry desert air is perhaps 99.5% air and 0.5% water. Ocean air is around 95% air and 5% water. And here's the pivotal fact: water vapour molecules are lighter than air molecules and exert less pressure on a barometer, an instrument designed to measure atmospheric pressure. Ipso facto, the approach of wet weather (fog, drizzle, rain, snow) is announced by a falling of the barometer.

The dials of most barometers are inscribed with weather terms: Fair, Change, Rain, and on older instruments, Moist, Dry, Stormy. Some are adorned with Greek gods seated on clouds, puffing against sailing boats. These are mostly meaningless, a memento from the weather lore days ('Red sky at night, sailors' delight'). Weatherwise, what mainly matters is the direction of the indicator's movement. Rising shows a change towards heavier, drier air; falling, a change towards lighter, moister air.

BARYON   See ELEMENTARY PARTICLES

BASE AND BASE PAIR (IN DNA)   See DNA AND RNA

BASIC   See COMPUTER LANGUAGES

BASIC RESEARCH   See SCIENTIFIC TERMS

BATTERY (CELL)

A source of direct current, usually produced by chemical changes. Strictly speaking, a battery is a group of cells connected together, but the term has come to be loosely applied to single cells as well as groups.

There are many kinds of batteries (cells), differing in size, shape, and components. All consist of two parts, called *electrodes,* made of two different substances; the electrodes are immersed in a liquid or paste (*electrolyte*), which reacts chemically with them. The two electrodes differ most importantly in how firmly they hold on to their electrons.

In an ordinary torch battery, shaped like a little can, one of the

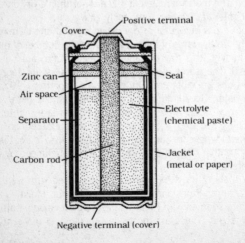

TORCH BATTERY

electrodes is the can itself, made of zinc. The other electrode is a carbon rod, in the centre. Zinc atoms have a looser hold on their electrons than carbon atoms do. When the zinc reacts with the electrolyte (a paste of ammonium chloride), its loosely held electrons are liberated, free to flow out, do their appointed work, and flow back into the carbon of the battery. When no more electrons are available, the battery is dead.

Some kinds of batteries can be recharged by connecting them to a source of direct current.

Car storage batteries are of this type, with the direct current coming from a generator driven by the car's engine (see ALTERNATOR). Some small batteries are also rechargeable. The most common are called *nicad batteries,* standing for nickel and cadmium, their electrode components.

You may have unwittingly experienced a natural battery in your mouth. Gold (fillings) and silver (fork) are the electrodes, and the slightly acid saliva is the electrolyte. The result is a tiny electric current, producing a sour taste and a slight shock.

Solar batteries are almost the ultimate in desirability: they never die and only temporarily fade away because there are no chemical changes. Light causes one of the electrodes to release its electrons, which flow out into the circuit (a radio, calculator, etc.) and then back to the other electrode, ready to be driven by light again and again. The stronger the light, the stronger the electric current. In ultrasunny Israel, several small solar cars are running around. They are energized by car-top solar batteries, whose current flows into electric motors that drive the wheels silently, with no stops for petrol — as long as the sun shines.

B-CELL *See* IMMUNE SYSTEM

BENIGN TUMOUR *See* CANCER

BERNOULLI EFFECT

A reduction of pressure in a fluid (i.e., a gas such as air, or a liquid) that is in motion. The faster the fluid moves, the more the pressure drops. Named after its discoverer, Swiss mathematician Daniel Bernoulli (1700–1782), the Bernoulli effect is measured and explained by the Bernoulli principle, which is full of equations of pressure, velocity, and conservation of energy and which can be found in appropriate reference books.

See also AERODYNAMICS.

BETA BLOCKER *See* ANGINA PECTORIS

BETA CELL   *See* DIABETES

BETA PARTICLE

An ELECTRON. Beta particles are emitted, among other forms of energy, by substances undergoing radioactive decay (*see* RADIOACTIVITY). Fast streams of beta particles are *beta rays* or *cathode rays*.

BETA RAYS   *See* BETA PARTICLES

BeV   *See* ELECTRON VOLT

BIG BANG

A theory of cosmology, based on the idea that some 15,000 million years ago all the matter and energy in the universe was concentrated in a single minute, hot, infinitely dense object. The explosion of this 'cosmic egg' blasted matter and energy in all directions in an expansion that continues today. The matter and energy of the cosmic egg are the matter and energy that make up today's stars and planets and all the other celestial objects of today's universe.

BILE ACIDS   *See* STEROIDS

BINARY DIGIT   *See* BIT

BINARY NUMBER SYSTEM

A numbering system used in computers, based on multiplying by two. Before examining the system, let's review the decimal number system, which is based on multiplying by ten (Latin *decem,* ten). In the number 77, the right-hand 7 means seven ones. The left-hand 7 means ten times as many, or 70 ones. Similarly, in the number 777, the left-hand 7 is worth ten times as many as the middle 7. Each move to the left multiplies the value of a number by ten. The number 307 means seven ones, no tens, and three hundreds. The decimal system uses ten numerals — 0, 1, 2, 3, 4, 5, 6, 7, 8, 9 — which can be used to construct any number, such as 369, 3.69, 3/69, or 0.00369. The idea of the decimal system, based on ten, probably originated in early humans counting on ten fingers (digits).

What if people had evolved with only two fingers? They would probably have developed a *binary* number system (*bi,* two). Fortunately for us and for Chopin, we did not so evolve. But the binary system did evolve, and it is ideally suited to computer operation (as we shall see in a moment). The binary system requires only two digits, 0 and 1. The value of a digit, as in the decimal system, depends on which column it's in. However, moving to the left multiplies the value of a digit by 2, not by 10. A few examples.

```
0001 one
0010 two
0011 three (two plus one)
0100 four
```

And so on, doubling to 8, 16, 32, and so forth as we continue to move left, column by column. Here's the number 13 in binary:

At first, the binary system may seem bulky and difficult to read — because it is. No doubt 11001000 seems a long-winded way to write *two hundred,* compared with 200. But there's one advantage that makes binary supremely suitable for computers: There are only two units, 0 and 1. Binary numbers can therefore be put in, stored, and transmitted by two positions of an electric switch: off (open) = 0 and on (closed) = 1.

Even the cheapest calculator contains several thousand switches in a little chip this size ☐, and a computer contains millions. Each switch can register a 0 or a 1. When you punch the number 5 on the keyboard, you cause a row of switches to set like this:

When you use such a calculator, you don't have to punch binary numbers into it, because there's a system of switches inside that converts decimal numbers into binary. The machine does the calculations in binary and converts the result into decimal to show on the face of the instrument.

See also INTERFACE.

BINARY STAR

When is a star not a star? When it's two or more stars. Stars look like single bright objects, but most of them are pairs. The partners revolve around each other, held in orbit by gravitation. The period (the time taken for one complete orbit) varies: less than a day to more than 100,000 years. Knowing the period, astronomers can calculate the mass (weight) of the partners. Pairs of stars are called *binary stars, binaries,* or *double stars.* There are also *multiple stars,* with trios, quartets, and larger groupings in orbit.

Binaries that can be seen with the naked eye (very few of these) or through a telescope are called *visual binaries.* Most binaries, however, are beyond the reach of even the most powerful telescopes but can be observed by joining a telescope to a spectroscope. Thus they are called *spectroscopic binaries.* As the partners in such a pair revolve around their common centre, the spectroscope differentiates between the approaching motion of one star and the receding motion of the other (*see* DOPPLER EFFECT).

The orbits of some binaries are so aligned that, from our viewpoint, star A eclipses (partly or wholly obscures the light of) star B. Later in the orbit, B eclipses A. This is an *eclipsing binary.* By comparing the varying amounts of light reaching the earth during a complete orbit of the pair, an astronomer can determine the brightness of each star and the orbit's inclination (angle) with respect to the earth.

BINDING ENERGY, NUCLEAR *See* NUCLEAR ENERGY

BIOCHEMISTRY *See* CHEMISTRY; MOLECULAR BIOLOGY

BIODEGRADABILITY

An organism lives, dies, and decays. In a relatively short time the large, complex molecules that make up its tissues are broken down (degraded) to form water, carbon dioxide, ammonia, and other substances whose molecules are small and simple. The agents of decay, mainly bacteria and moulds, are themselves alive, hence the term *bio*degradability — the capability of being broken down by the action of living things. The wastes of organisms — their urine and faeces, for example — are also biodegradable, as are materials made from organisms — paper from trees, cloth from cotton, leather from animal hides, and the like.

The decay process produces the small, simple molecules from which the tissues of new organisms are built. The new organisms in turn die and decay in a natural recycling process.

Some artificial materials are degradable — slowly — by nonliving agents. Steel, for example, rusts by combining with oxygen in the air. But some synthetic materials, especially plastics, are virtually degradation-proof. The growing use of plastics in bottles and for sizable parts of cars, for example, is causing an increasingly serious problem — how to dispose of used plastic. At present, the recycling technology is available but unprofitable.

BIOFEEDBACK

A technique whereby people learn to control *autonomic* (ordinarily non-controllable) functions of the body, with the help of electronic equipment. The technique is based on the idea that a person made aware of what is happening with an autonomic function — the blood pressure, for example — can develop control over it.

An individual, say one with high blood pressure, is connected to a device that monitors the blood pressure continuously; the device shows changes in the pressure by changing the brightness of a light or the pitch of a tone. The subject concentrates on changing the light or the tone by trying to lower the blood pressure. Even a very small change is immediately apparent. Encouraged by small successes, the subject tries for bigger ones. In time he may establish enough control to do without the machine.

Experimenters have succeeded in controlling sweating, warming their hands or feet, controlling the rate and regularity of the heartbeat, and other autonomic functions. Proponents of feedback therapy believe that someday it will be used widely as a basic medical technique.

BIOLOGY

The science that deals with living things (Greek *bios*, life). Formerly broadly divided into two areas, zoology (Greek *zoon*, animal), the study of animals, and botany (Greek *botanes*, plant), the study of plants, biology is now divided and subdivided into hundreds of special fields involving the structure, function, and classification of the forms of life from the simplest, the viruses (Latin *virus*, poison), to the most complex, *Homo sapiens*. Some of the more important divisions are:

Anatomy (Greek *anatome*, dissection) The study of body structure of plants and animals. The study of structural similarities and differences between related forms of life is called comparative anatomy.

Cytology (Greek *kutos*, hollow vessel) The study of cells. A very broad field, considering that over 99% of all living forms are made of cells. However, cytology deals more specifically with the individual cell as the unit of study. Thus: how does a cell in the root of a marine plant accept water but reject salt?

Ecology (Greek *oikos*, house) The scientific study of the relationship between organisms and their environment.

Embryology (Greek *embroun,* something that grows in the body) The science that deals with the formation, early growth, and development of living organisms.

Evolution (Latin *evolutio,* opening, unrolling) The branch of biological science dealing with the theory that existing forms of life developed by a process of continuous change from previously existing forms. The mechanisms of these changes are embodied in genetics.

Genetics (Greek *genesis,* origin, creation) The study of heredity, especially the study of the biological mechanisms by which characteristics vary and are transmitted.

Histology (Greek *histos,* web) The microscopic study of tissues, the sheets of cells that together make up skin, muscle, bark, fat, petals, and so on.

Palaeontology (Greek *palai,* long ago being) The study of fossils and ancient forms of life. Further divisible into palaeobotany, palaeozoology, palaeoanthropology (the scientific study of the origin and of the physical, social, and cultural development of ancient man), and so on.

Physiology (Greek *phuein,* to make grow) The study of the essential life processes, such as digestion, circulation, photosynthesis, metabolism, and so on.

Taxonomy (Greek *taxis,* arrangement, order) The study of the system by which living things are classified in established categories such as phylum, order, family, genus, and species.

See also MOLECULAR BIOLOGY.

BIOMASS

A general term for organic material, both living and nonliving.
See also ENERGY RESOURCES.

BIOPHYSICS *See* PHYSICS

BIOTECHNOLOGY

The industrial use of microorganisms and living plant and animal cells to produce substances or effects beneficial to people. Biotechnology encompasses the manufacture of antibiotics, vitamins, vaccines, plastics, and feedstuffs; TOXIC-WASTE DISPOSAL using bacteria; pollution control; the production of new fuels; and much more. Future development will rely heavily on GENETIC ENGINEERING.

BIT

Contraction of *binary digit*. The smallest unit of information in the binary system of notation, a bit can be either 1 or 0, referring to the two positions of a switch; closed or open. Thus the binary number 1001 (number 9 in the decimal system) is indicated by four switches, two of which are in the closed position (the ones) and two in the open position (the zeros).

Computers need this two-digit kind of simplicity because they work by switches that can be in only one of two conditions, on or off. However, by using groups of bits, called *bytes,* arranged according to various systems, they can convert letters, numbers, and symbols (such as @, $, [S#], &) into binary form. For each character on a binary computer keyboard, there is one and only one byte that represents it. For example, in the ASCII code the byte size is 7 bits. The first bit (at the left) is a yes or no statement about whether the whole byte is a letter of the alphabet (yes is 1, no is 0). Here is the byte for capital A: 1000001. The byte for number 1 is 0110001. The left 0 is clear, and so is the right 1; the other bits have to do with other aspects of the ASCII code, whose explanation is too space-consuming for this book.

See also BINARY NUMBER SYSTEM.

BLACK BOX

A device, or an idea of a device, whose actual details are unknown but whose function within a larger system is known. 'When I turn the ignition key, something makes the engine start' is a black-box statement because it acknowledges the existence of an intervening 'something' without describing the functions of a car battery, solenoid switch, Bendix drive, and starter motor. On the other hand, the statement 'When I turn the ignition key, I make the engine start' is *not* a black-box statement.

Scientists and engineers designing a highly complex project, such as a communications satellite, usually isolate and postpone the portions of the design that are most likely to be solved as black boxes in order to concentrate on the pitfall aspects that could endanger the entire project.

BLACK DWARF STAR   *See* STELLAR EVOLUTION

BLACK HOLE   *See* STELLAR EVOLUTION

BLOOD PRESSURE

The pressure exerted by the blood against the walls of the blood vessels. A blood pressure reading is an important part of every physical examination, since it gives the doctor insight into the condition of the heart, blood vessels, and other organs of the body.

Two pressures are actually recorded, usually at the artery in the upper arm:

- Systolic (Greek *sustellein,* to draw together) is the pressure recorded when the heart contracts, squeezing the blood.
- Diastolic (Greek *diastellein,* to be dilated) is the pressure recorded when the heart stops squeezing and relaxes.

Blood pressure is measured with a pressure gauge called a *sphygmomanometer.* The most common form of this instrument employs a column of mercury, the height of which is graduated in millimetres. Thus 120/80 indicates a systolic pressure of 120 mm of mercury (mm/Hg) and a diastolic pressure of 80 mm.

BLUE SHIFT   See DOPPLER EFFECT

BOILING-WATER REACTOR   See NUCLEAR REACTOR

BOLIDE   See SOLAR SYSTEM

BOOLEAN ALGEBRA

George Boole (1815–1864), professor of logic and mathematics at Queens College, Cork, would be surprised to arise and discover that a multimillion-pound worldwide industry is based on his thinking, especially as expressed in one of his books, *An Investigation of the Laws of Thought,* published in 1854. Boole asserted that most logical thinking, cleared of fluff and verbiage, can be conceived as a series of choices. Boole's interesting notion became the basis of electronic computers of all kinds. The programming for little pocket calculators, home computers, arcade games, and even the computers for interplanetary space vehicles is based on Boole's principles of logical thought, now called Boolean algebra. This algebra, in turn, is based on the three little words AND, OR, and NOT, and their combinations, NOT-AND (NAND) and NOT-OR (NOR).

But how does human-generated logic get into a computer, much less accomplish anything worthwhile there? The ideas represented by AND, OR, NOT, and their combinations can be imitated by electric circuits controlled by on-and-off switches. First, let's look at

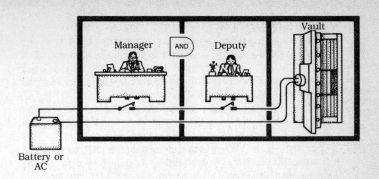

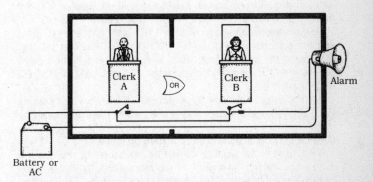

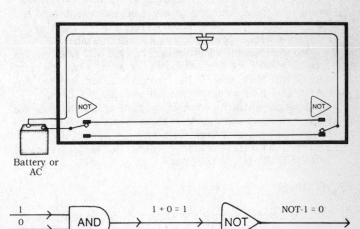

$\dfrac{1}{0}$ → AND Gate → $1 + 0 = 1$ → NOT Gate → NOT-1 = 0 →

a noncomputer example of an AND circuit, the electrically controlled lock of a bank vault door. To open the vault, a switch must be turned on by the bank manager AND another by the deputy manager, each at a secret place in his or her office. In a computer, a CHIP the size of your fingernail may contain up to 200,000 electronic switches called *gates* that perform the AND function.

Back to the bank. It has a holdup alarm system, which can be switched on by bank clerk A OR bank clerk B. A tiny computer chip may have 200,000 OR gates in addition to the AND gates.

Once again to the bank, whose files are kept in a long, narrow room with a door at each end. The overhead light can be turned on or off by a switch near each door. This is an example of a NOT gate, because either switch, when operated, makes the light do whatever it is *not* doing. Our computer chip has a mere 20,000 or so NOT gates.

Only two more gates to go. Attach a NOT gate past an AND gate, and the two become a NOT-AND (NAND) gate. Whatever the first gate does, the second gate, when operated, reverses the action. Similarly, a NOT gate connected past an OR gate makes a NOT-OR (NOR) gate.

You could figure out how the switches at the bank work by taking them apart. But what happens on a chip, where there are no moving parts? Just looking, even through a microscope, tells you very little, because the chip works on a very different principle.

Well, what do the switches do that we can't observe? One example is in a calculator, which can't do subtraction directly. But a way of getting around this disability is built into the machine. This involves, among other steps, changing zeros to ones and ones to zeros and then adding. In the BINARY NUMBER SYSTEM, all numbers can be made out of zeros and ones, and the calculator does this.

When you press the minus key, you connect NOT gates into the circuits, making the ones NOT-ones, or zeros, while the zeros become NOT-zeros, or ones. Then the machine adds and performs the final step of converting the binary answer back to decimal.

A more complicated example is the game of draughts, where a player, human or electronic, makes choices that are really Boolean in nature. The grandest strategic decisions in the game of draughts, or even chess, are made up of little yes-or-no steps that can be imitated by the on or off position of switches.

BOSONS

Certain subatomic particles that transmit forces. Among these are photons (which transmit electromagnetic forces) and gravitons (which transmit gravity).

See also ELEMENTARY PARTICLES.

BRAIN HAEMORRHAGE See STROKE

BRAIN WAVES See ELECTROENCEPHALOGRAPH

BRAND-NAME DRUGS See GENETIC DRUGS

BRASS See ALLOY

BREEDER REACTOR See NUCLEAR REACTOR

BRONCHI See EMPHYSEMA AND BRONCHITIS

BRONCHITIS See EMPHYSEMA AND BRONCHITIS

BUBBLE CHAMBER See PARTICLE ACCELERATOR

BULIMIA See ANOREXIA NERVOSA AND BULIMIA

BURKITT'S LYMPHOMA See HERPESVIRUS DISEASES

BYTE See BIT

CALCULATOR

The common hand-held or desktop device that performs arithmetic operations in a fixed, step-by-step order, with each step requiring a specific instruction by an operator. By contrast, a computer can be given a whole series of instructions, called a program, to perform in sequence.

See also BOOLEAN ALGEBRA; COMPUTER.

CALCULUS

Mathematics on the wing: its most subtle, imaginative, and powerful discipline. By contrast, the more familiar branches — arithmetic, algebra, and geometry — might be regarded as mathematics with its feet on the ground. What was the average speed on our journey? (arithmetic). If we hire two workers in addition to the three already at work, how much sooner will the job be done? (algebra). How many square metres of shingles are needed to cover the house, with its gables, dormers, and angles? (geometry). All these examples deal with fixed, stable quantities: distance and time, workdays and workers, square feet and polygonal dormers; all are definite and dependable, good citizens.

Now consider this seemingly simple variation. A sack of sand is suspended, with a small hole at the bottom from which sand trickles at a constant rate (e.g., 1 cubic centimetre per second). The sand forms a small, cone-shaped hill that grows in height, diameter, volume, and weight. What will these measurements be, *and at what rate will they be changing,* after three seconds? five seconds? A single constant input (1 cubic centimetre per second) has produced several outputs; is each output changing at the same rate? Furthermore, the input is *not* constant, because as the sack gradually becomes emptier and its load lighter, the sand trickle gradually becomes slower. So with a varying input, will the *rate of change* of each output vary in its rate of variation?

Precisely this kind of numbing question must be solved by the designers and engineers of a great variety of machines, instruments, and structures. Take a spectacular example: designing a spaceship. To achieve sufficient speed (escape velocity) to overcome the effects of the earth's gravity requires juggling the following variables (and more):

- The more powerful the engines, the faster the spaceship climbs.
- But more powerful engines are heavier and consume more fuel, at a faster rate.
- Consuming more fuel means carrying a heavier total weight at takeoff.
- And let's not forget that a heavier total weight requires more power to get it off the ground.

Somewhere in that quartet of variables (and others unmentioned), among literally billions of possibilities, there is hidden the ideal choice: the right fuel capacity, engine power, and fuel flow to launch the spaceship most efficiently, with maximum reserve power. Can the answer be found by precalculus means? Yes, in perhaps 200 worker-years of algebra. By using calculus? A few days. Using calculus programmed into a computer? A few minutes.

The essence of calculus is in its concept of change. The mathematician and philosopher Bertrand Russell (1872–1970) put it this way: 'People used to think that when a thing changes, it must be in a state of change, and that when a thing moves, it is in a state of motion. This is now known to be a mistake'.

Then what is not a mistake? Films are not a mistake. TV is not a mistake. In these visual depictions and perceptions, motion is a series of still pictures in rapid sequence: 24 per second in films, 25 per second on TV. So motion might be regarded for convenience as an infinite number of still pictures per second. Calculus consists of dividing all changes (increase in diameter of sand pile, decrease in weight of fuel, gain of altitude in rocket, etc.) into an infinite series of stills without actually doing all that infinite work and then pushing those stills around. There are two kinds of pushing:

1. *Differential calculus.* Differential calculus involves finding the rate at which something is happening (or is changing in its rate of happening) in reference to a related happening. How much taller (per second) is the sand pile growing when the sand is spilling out at half its initial rate? (Substitute a few space-launch terms if you're tired of sand piles.)
2. *Integral calculus.* Integral calculus involves knowing the rate at which something is happening and finding the various combinations of happenings that could possibly be producing it. The rocket engine's horsepower, on the test stand, is gaining at the rate of 10% per minute. What various rates of fuel consumption in relation to oxygen consumption can be producing that particular rise in power?

No, your little home calculator will not do calculus for you. But an elementary high school text is worth a try, just to get a rough idea of the potency of calculus. And you will probably enjoy reading about the two independent, almost simultaneous inventions of calculus, by Isaac Newton and Gottfried Wilhelm von Leibniz.

CANCER

A group of more than 100 diseases, second only to heart disease as a cause of death in both the United Kingdom and the United States. In the UK, cancer accounts for approximately 25% of all deaths compared with the worldwide figure of 8%. The incidence of both cancer and heart disease has increased sharply in the western world this century, mainly because the virtual conquest of infectious diseases has enabled many more people to reach middle and old age, when susceptibility to cancer and heart disease is greatest.

Normally, the body cells — skin, muscle, bone, and so on — grow and divide in an orderly way, at a controlled rate, producing more cells exactly like themselves. This is how we replace old tissues and grow. Sometimes disorganized, uncontrolled division occurs, and a tumour forms (Latin *tumere,* to swell) (also called a *neoplasm,* from the Greek *neo,* new + *plasma,* form). Tumours that grow slowly without spreading, such as warts and moles, are called *benign* (Latin *benignus,* well-born). They are not cancerous, but in some cases the pressure exerted by a benign growth (a tumour of the brain, for example) may be dangerous.

In *malignant* (cancerous) tumours, some of the cells may spread to neighbouring organs (infiltration); they can also enter the bloodstream and be carried to other parts of the body, starting new tumours there. This colonization process is called *metastasis* (Greek *methistanai,* to change).

The many forms of cancer are classified into four groups:

1. *Carcinomas,* from the Greek *karkinos,* 'crab', after the fancied clawlike spread of the disease; similarly, cancer is Latin for 'crab'. Carcinomas involve the skin and the skinlike membranes (epithelium) of the internal organs.
2. *Sarcomas* (Greek *sarkoma,* fleshy growth). They involve the bones, muscles, cartilage, and fat.
3. *Leukaemias* involve the white blood cells, or *leucocytes* (Greek *leukos,* 'clear, white' + *kytos,* hollow vessel, cell).
4. *Lymphomas* involve the *lymphatic system,* the network of vessels and tissues that recaptures lymph (Latin, water), the colourless blood fluid that seeps out of the tissues.

There are many *carcinogens,* agents with a potential for producing cancer, including tobacco smoking, smoke from industry, X-rays, certain dyes, asbestos, nuclear radiation, and the ultraviolet rays in sunlight. Their link to cancer was long known, yet there was a mystery: Could such a wide variety of agents be the cause? Scientists theorized a two-step process: (1) Carcinogens cause changes (mutations) in the genes within a cell, and (2) the changed genes cause the cancer.

Genes (see DNA AND RNA) are the hereditary material that determines the characteristics of every living thing. When a cell divides in two, each 'daughter' cell receives an exact copy of the genes, organized into *chromosomes* — rod-shaped bodies of DNA molecules.

CANCER IMMUNOLOGY   See CANCER

CAPACITOR, VARIABLE   See RADIO

CAPILLARY   See CIRCULATORY SYSTEM

CARBON DATING   See UNCERTAINTY PRINCIPLE

CARBON 14   See UNCERTAINTY PRINCIPLE

CARCINOGEN   See CANCER

CARCINOMA   See CANCER

CARDIAC PACEMAKER

A bundle of nervous and muscle tissue in the upper wall of the heart that regulates the heartbeat rate. The *sinoatrial node,* as it is technically known, generates electrical impulses at a regular rate — about 70 times per minute in adults, rising to 150 or more during exertion. Each impulse spreads across the *atria* — the two upper chambers of the heart — causing a contraction that forces the blood within them into the *ventricles,* the pair of muscular chambers that make up the bottom part of the heart. A second bundle of tissue between the chambers passes the impulse along, causing the ventricles to contract and forcing the blood into arteries that lead it away from the heart.

Several kinds of heart disease may affect the pacemaker, resulting in a heartbeat that is too fast, too slow, or irregular. The remedy for these disabling or life-threatening conditions is often the widely used *artificial pacemaker,* a device that is implanted under the skin and connected to electrodes permanently implanted in the heart muscle.

Powered by long-lasting batteries, the device generates electrical impulses similar to those produced by the natural pacemaker.
See also CIRCULATORY SYSTEM.

CARDIAC SURGERY See OPEN-HEART SURGERY

CARRIER WAVES See RADIO

CASTING (METAL OR PLASTIC)
See MANUFACTURING PROCESSES

CATALYSIS AND CATALYSTS

Behold a chemist, directing the actions of billions of molecules in a flask. They combine, separate, change partners – reacting exactly as ordered. If a reaction doesn't go fast enough, or at all, forceful methods of persuasion are available: intense heat, high pressure, strong acids. There is also a gentler method, catalysis.

A catalyst is a substance with special characteristics:

- Very little of it is needed.
- It is intimately involved in the reaction but is not itself changed; therefore, it isn't used up.
- It is specific – that is, it catalyses only one reaction or a group of closely related reactions.

An example of catalysis is the operation of the *catalytic converter* fitted in the exhaust systems of cars that run on unleaded petrol. One of the catalysts is a mixture of platinum and palladium, two of the most precious of the precious metals. Fortunately, a tiny amount is all that's needed, and it isn't used up. It speeds up the reaction that combines unburnt fuel (hydrocarbons) and poisonous carbon monoxide from the exhaust with oxygen from the air to form water and carbon dioxide. At the same time, a platinum and rhodium catalyst helps to remove oxygen atoms from the air-polluting nitrogen oxides in exhaust fumes (see ACID RAIN), forming harmless nitrogen and oxygen.

Other examples of catalysis:

- Powdered nickel catalyses the reaction in which liquid vegetable oils are changed to solid fat.
- 'Cat-cracking' (*cat* is short for *catalytic*) is one of the principal steps in oil refining. The heavier ingredients in petroleum, such as tar and asphalt, are split up into lighter molecules, to form petrol,

paraffin, flavours (yes!), dyes, aromatic substances, and numerous other chemical triumphs.
- The thousands of chemical reactions that take place in living things are promoted by catalysts called *enzymes.*

CATALYTIC CONVERTER
See CATALYSIS AND CATALYSTS

CATARACT

A condition affecting the lens of the eye, mainly in older people. Worldwide, there are approximately 17 million cases. In the United States, cataract is the leading cause of blindness after diabetes. Like a camera lens that focuses light rays to form an image on film, the eye lens forms an image on the retina at the back of the eyeball. The lens is normally transparent, but in the cataract condition it becomes cloudy; like a mist-covered window, it diffuses the image. The degree of cloudiness may be so slight as to escape the patient's notice or heavy enough to produce blindness.

The cause of cataract is unknown, and medicines are ineffective against it. However, cataract operations are a triumph of modern surgery, with a success rate of more than 90%. The painless operation involves removal of the clouded lens and substitution of a functional equivalent. At one time, this involved thick-lensed spectacles that offered only limited vision. Later, contact lenses produced much better results but were often impossible for older people to use. Today, *lens implants* are used in many of the operations, especially those on older patients. The natural lens is replaced by a tiny plastic lens. For most patients this restores vision that is very close to normal, although spectacles are usually necessary for reading.

'CAT-CRACKING' See CATALYSIS AND CATALYSTS

CATHODE RAY TUBE (CRT)

The most familiar example is a television picture tube, and the simplest kind is the black and white. The inside face of the tube is coated with a *phosphor,* a substance that glows when struck by electrons. At the rear of the tube (the neck) is an electron gun that shoots a beam of electrons towards the front. Electromagnetic coils or electrically charged metal plates direct this stream from side to side and top to bottom, forming a glow-picture of the 'message' being received by the cathode ray tube. Colour tubes are similar except that the face is coated with thousands of groups of dots of light-emitting phosphor. Each group, called a *pixel* (picture element), consists of three dots, one for each of the three primary colours — red, green, and blue.

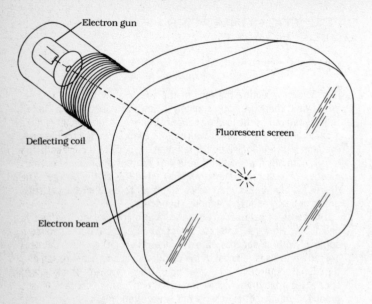

CAT SCAN *See* COMPUTERIZED AXIAL TOMOGRAPHY

CD *See* SOUND RECORDING

CELESTIAL COORDINATES

Think of the problem you may face some enchanted evening if you happen upon a UFO and wish to report its precise location. Celestial coordinates are the answer. Imagine, as the ancients once did, that the earth is surrounded by a huge crystal globe, a *celestial sphere,* on which the stars have been fixed in place. There is, of course, no such sphere, but the idea, in combination with a set of imaginary lines, is useful for specifying the location of a star, a planet, or even a UFO.

To locate an object in the sky, you must have some way of knowing how far 'up' it is and how far to the right or left of some given starting point. To do this, we apply the method used on the spherical earth: a system of coordinates — parallels of latitude and meridians of longitude.

The earth's parallels of *latitude* begin at the equator. North of that is the 1st parallel north, and so on up to the 90th parallel north at the North Pole. A similar set of parallels lies to the south of the equator.

The meridians of *longitude,* 360 of them, radiate from the North Pole, cross the equator, and meet again at the South Pole. The meridian that passes through Greenwich is the *prime meridian.* The meridian to its west is 1 degree west longitude, followed by 2 degrees west longitude, and so on, until, halfway around the earth, we reach 180 degrees west longitude. A similar set of meridians stretches eastward from Greenwich, ending in 180 degrees east longitude (which is the same meridian as 180 degrees west longitude).

Now imagine the earth's coordinates projected outwards onto the celestial sphere. There they form a system of coordinates that can be used for locating stars, planets, galaxies, and UFOs. The nomenclature is a bit different. Latitude is called *declination,* and longitude enjoys the name of *right ascension.*

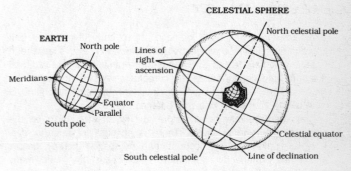

CELESTIAL SPHERE   See CELESTIAL COORDINATES

CELL CULTURE   See TISSUE CULTURE

CELL (ELECTRIC)   See BATTERY (CELL)

CENTRAL PROCESSING UNIT (CPU)

Could be called the brain of a COMPUTER, if computers had brains. The CPU carries out the computer's arithmetic, logic, and control operations, as this simple example indicates:

> Record the following exam grades for each student on this list. Then average the exam grade into the student's previous

average. If the averages are the same, or within five points, print the student's name on List A. If the new average is more than five points above the previous average, print the student's name on List B. If more than five points below, print on List C.

Arithmetic operations were performed to obtain the averages. Logic operations were required to compare the new averages with the previous ones. (Are these numbers the same? If not, which is larger?) Control operations caused a PRINTER — a kind of electric typewriter — to print the three lists.

CENTRIFUGAL CASTING
See MANUFACTURING PROCESSES

CENTRIFUGATION *See* ANALYSIS

CEPHALOSPORINS *See* ANTIBIOTICS

CEREBRAL APOPLEXY *See* STROKE

CEREBRAL PALSY

A disorder in which brain damage affects muscular control and coordination. There may be mental retardation, and speech and hearing may be affected. Impairment may range from slight to total disability and dependence. The cerebral palsy patient may suffer from sudden muscular contractions, spontaneous disorganized movements, trembling, or loss of balance.

Brain damage may occur before, during, or after birth, as the result of infection or injury. There is no cure for the disorder. However, physiotherapy, occupational and speech therapy, braces and other orthopaedic appliances, and counselling help many patients to lead productive lives.

CEREBROVASCULAR ACCIDENT *See* STROKE

CERN *See* PARTICLE ACCELERATOR

CFC *See* OZONE LAYER

CHAIN REACTION *See* NUCLEAR ENERGY

CHEMICAL ENERGY *See* ENERGY

CHEMICAL ENGINEERING *See* CHEMISTRY

CHEMICAL TESTS See ANALYSIS

CHEMISTRY

The term *chemistry* can be traced backwards in time and space. Through Middle English, French, Greek, and Arabic branches, we reach roots it shares with the term *alchemy,* which is the name for an ancient art of unknown origin that sought to transmute base metals like lead into gold and silver. Alchemy was a forerunner of the modern science of chemistry, which deals with the composition, structure, properties, and reactions of matter, especially at the atomic and molecular levels.

Like all basic sciences, chemistry has become divided into numerous areas; former sharp boundaries have become broad zones.

Analytical chemistry The science and techniques by which the chemist determines the answers to the questions 'What ingredients are in this substance?' and 'How much of each ingredient is there?' (*see* ANALYSIS).

Biochemistry The study of the chemical processes that go on in living things — in short, the chemistry of life.

Electrochemistry The science of the inter-relationships of electric currents and chemical changes, including such processes and devices as electrical batteries (cells), metal refining and manufacture, the production of hydrogen, chlorine, and other chemicals, and metal plating.

Geochemistry The study of the chemical composition of the earth's crust, waters, and atmosphere. Among the practical aspects of this science are the location of mineral ores, natural gas, and petroleum.

Polymer chemistry The science of giant, chainlike molecules, *polymers,* made by the repeated linking of great numbers of simple molecules called *monomers.* Rubber and cellulose are natural polymers. Artificial rubber, plastics, and nylon are artificial, or synthetic, polymers.

Industrial chemistry The business aspect of chemistry, the application of chemical science to the production of fuels, products, and by-products. Closely related to *chemical engineering.*

Physical chemistry A bridging area of science encompassing the way that the physical properties (weight, volume, hardness, etc.) of a substance depend on its chemical composition and what physical changes accompany a chemical change.

Organic chemistry Formerly defined as the chemistry of living matter, now the chemistry of carbon compounds.

Inorganic chemistry Broadly speaking, the chemistry of all substances not containing carbon.

Here are some more terms describing specialized fields of interest in chemistry: *pharmaceutical chemistry* – medicinal drugs; *structural chemistry* – atomic and molecular arrangements and linkages; *thermochemistry* – energy transfers, especially of heat, during chemical reactions.

See also ELEMENTARY PARTICLES; PROTEINS AND LIFE.

CHEMOTHERAPY See CANCER

CHICKEN POX See HERPESVIRUS DISEASES

CHIP

A base (substrate), usually of silicon, on which a group of electronic circuits are constructed. Most chips measure less than 1 square centimetre (0.155 square inch) in area and contain hundreds of thousands of parts. Most of the parts are tiny switches (transistors) that process or store information. The parts are not assembled on the substrate but are formed out of it. The whole assemblage, called an *integrated circuit,* is roughly like a photographic image formed out of a photographic emulsion. The original of the image is a large diagram that is reduced by a reducing lens – the opposite of an enlarging lens. The exposed reduced image is then processed by etching, plating, and photographic processes.

See also COMPUTER; SEMICONDUCTOR.

CHLOROFLUOROCARBONS See OZONE LAYER

CHOLESTEROL

A fatty substance produced by the body, mainly in the liver and intestine, and also ingested in foods of animal origin, such as butter, eggs, and fatty meats. Cholesterol belongs to a large group of compounds called the STEROIDS. Its functions of cholesterol are not fully understood, but some facts are known: it is present throughout the body, it is especially abundant in the nervous system, and it is the raw material of certain hormones, bile salts, and the membranes of the body cells.

It is clearly a necessary substance, yet its name provokes unease. That's because cholesterol may become too much of a good thing if an excessive amount circulates in the bloodstream. Sometimes, for reasons unknown, deposits, called plaques, of cholesterol form on the inner lining of arteries (see ARTERIOSCLEROSIS), hindering the blood flow and possibly causing blood clots. *Triglycerides,* a group of fatlike substances, and the form in which fat is stored in the body, may also be involved. In arteries of the heart, brain, or kidneys, the end result may be a heart attack, stroke, or kidney failure.

To lower the level of cholesterol in the blood, most doctors look first to the patient's diet, advising a lower intake of foods rich in cholesterol and saturated fats and an increase in consumption of fish and fibre. Weight loss, regular exercise, and stopping cigarette smoking are recommended. Drugs may be prescribed.

The molecules of a fat or oil consist of chains of carbon atoms, to which pairs of hydrogen atoms are attached. A fat is *saturated* if its chains are fully loaded with hydrogen, as in this partial structural formula:

Butter is such a fat.

Unsaturated fats are considered less likely to raise the cholesterol level of the blood. There are degrees of unsaturation. If the chain lacks *one* pair of hydrogen atoms, the fat is said to be *mono*unsaturated. If *more* than one pair is missing, the fat is *poly*unsaturated.

Peanut oil, sunflower oil, and most other vegetable oils contain relatively high amounts of unsaturated molecules. However, they are often saturated by the addition of hydrogen. This *hydrogenation* process is used to improve the flavour or odour of the oil or to convert it to a solid, such as margarine.

Doctors are interested in the relative amounts of high-density lipoprotein (HDL) cholesterol and low-density lipoprotein (LDL) in the blood. HDL is sometimes called the 'good cholesterol' because it is believed to facilitate the removal from the body of the potentially harmful LDL. Regular exercise and polyunsaturated fats appear to increase the HDL level.

Drugs can help in lowering the level of cholesterol. In 1987, an especially promising drug, *lovastatin,* received approval by the Food and Drug Administration in the United States. It is an enzyme inhibitor, interfering with the production of LDL cholesterol by the liver.

See also PROTEINS AND LIFE.

CHOLINESTERASE INHIBITORS
See NEUROMUSCULAR DISORDERS

CHORIONIC VILLUS SAMPLING   See AMNIOCENTESIS

CHROMATOGRAPHY   See ANALYSIS

CHROMOSOMES   See DNA AND RNA

CHROMOSPHERE   See SOLAR SYSTEM

CIRCUIT BREAKER

A device that does the job of a FUSE without self-destructing. Current flows through a switch and an electromagnet (A). The strength of the electromagnet depends on the strength of the current flowing through. Too much current causes the electromagnet to become too strong, so strong that it pulls the switch to an off position, breaking the circuit (B). After the overload has been corrected, the switch is closed manually and the current restored. Some circuit breakers operate by the heating of a metal strip. A current overload causes the strip to heat up and bend, thus tripping the switch to the 'off' position.

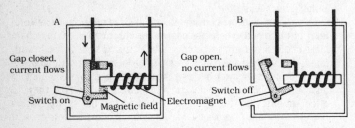

CIRCULATORY SYSTEM

The system of 'pumps' and 'pipes' that distributes the blood throughout the body. The pump is the *heart,* which is actually a pair of pumps. The right side of the heart pumps blood directly to the lungs; the left side pumps blood to the rest of the body through various pathways:

Arteries The large pipes that carry the blood away from the heart. They extend to all parts of the body, where they branch to form networks of smaller pipes, or *arterioles;* these in turn branch to form capillaries.

CIRCULATORY SYSTEM

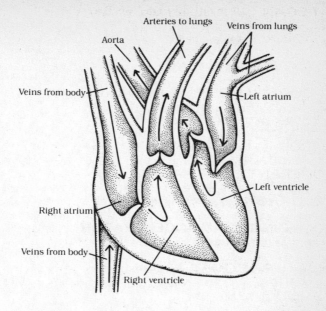

Capillaries Microscopically thin blood vessels intimately entwined with the body cells. Here, dissolved food, oxygen, hormones, and hundreds of other substances pass out of the blood into the body cells; wastes and many other substances pass out of the body cells into the blood. The capillaries merge to form *venules,* which join to form veins.

Veins Larger blood vessels that return the blood to the right side of the heart. The heart pumps this blood to the lungs for aeration, then back to the left side of the heart, where it is pumped out on another trip around the body.

In the heart, each contraction of the left side forces a quantity of blood (about 5 tablespoons) into the *aorta,* the largest artery. Like all arteries, it is elastic and muscular. It stretches, then contracts, pushing the blood along and setting off a wave of stretching and contracting in the arteries. In places where an artery can be pressed against a bone — at the wrists, ankles, or temples, for example — the motion is felt as the *pulse*.

The pulse indicates the rate, regularity, and strength of the heartbeat. Together with the other *vital signs* — body temperature, respiratory rate, and BLOOD PRESSURE — it offers the medically trained person numerous clues to the body's state of health.

CIT SCAN See COMPUTERIZED AXIAL TOMOGRAPHY

CLADDING See FIBRE OPTICS

CLONES See IMMUNE SYSTEM

CLOUD CHAMBER See PARTICLE ACCELERATOR

COBOL See COMPUTER LANGUAGES

CODON See DNA AND RNA

COHERENT LIGHT See LASER

COLD FRONT See AIR MASS ANALYSIS

COLLIDING-BEAM ACCELERATOR
See PARTICLE ACCELERATOR

COMA OF A COMET See SOLAR SYSTEM

COMET See SOLAR SYSTEM

COMFORT INDEX See TEMPERATURE-HUMIDITY INDEX

COMMUNICATIONS SATELLITE See SPACE TRAVEL

COMMUTATOR See GENERATORS AND MOTORS, ELECTRIC

COMPACT DISC See SOUND RECORDING

COMPOUND MICROSCOPE See MICROSCOPE

COMPUTER

Device used for computation, as for example your fingers, if you like to count on your fingers. More precisely, the contemporary definition (and doubtless the one you are looking for) refers to an electronic device that (1) receives *input,* (2) *stores* and *processes* that input according to a set of instructions (a *program*), and (3) delivers the results of that processing in the form of an *output*.

A shop assistant lays a bunch of bananas on a computer-equipped weighing scale. The weight of the bananas acts as input 1, which is temporarily stored awaiting input 2, the price per pound of bananas, which is punched in by the assistant. The computer processes the two inputs according to a built-in program: 'Multiply inputs 1 and 2'. The result is an output, in the form of lighted numbers — pounds and pence — on the dial of the scale.

The captain of a 440-passenger plane, after take-off from Heathrow, punches a set of data (input) into the keyboard of a computer-equipped automatic pilot, indicating the desired speed, altitude, and flight path for a flight to Rome. The data are stored in the computer's memory, awaiting a second set of inputs, which are continually received throughout the flight from the air speed meter, the altimeter, the compass, and other instruments. Processing these several inputs, the computer delivers a continual series of outputs — instructions to the engine controls and to the rudder and elevator — producing a smooth, hands-off flight all the way to Rome, right through the landing, if necessary.

Input, storage (memory), processing, and output — these are the basic steps in the banana transaction and in the flight to Rome. The same basic steps apply in personal microcomputers; in commercial (*mainframe*) computers the size of a car; in laptop and desktop computers; in pocket calculators, which are a kind of simple computer; and in fact in all computer operations. Obviously these differ in complexity, but the machinery involved, the *hardware* units, are surprisingly similar, and many of them are even interchangeable. Here are descriptions of some typical operating segments.

Most computers are equipped with typewriterlike keyboards (keypads). Electric switches under the keys convert letters, numbers, and symbols (punctuation marks, dollar sign, etc.) into binary digits or BITS — zeros and ones — in the ASCII code. A place where two systems make contact is called an INTERFACE. In this case the two systems are the ALPHANUMERIC system and the BINARY NUMBER system.

Input Inside the weighing scale, input about the weight of the bananas is converted at an interface into binary form. On the aeroplane, continuous messages about air speed, altitude, and compass heading are converted by TRANSDUCERS into electric currents that reach interfaces in the computer and are then converted into binary. Another common input is the UPC (universal product code), also called the bar code, seen on packages. Here the interface is on a laser-beam scanner. It converts the message of thick and thin lines into binary information for the computer. Similarly, a MAGNETIC INK CHARACTER SORTER 'reads' the numbers, printed with magnetic ink, on cheques and other documents. Many post offices have optical readers that read post codes (especially if they're written neatly). In banks with machine-operated cash dispensers, two inputs are needed for identification: a magnetically inscribed code number and a private number that the customer must punch into the keyboard.

Another type of input is punched cards or tapes. A sales assistant snips off a punched card attached to the coat you've just purchased and puts it into a device that 'feels' the punched-out holes, with either metal fingers or light beams. The coded information (manufacturer, price, size, etc.) is converted to binary language, stored in the computer's memory, and sent to the cash register.

There are many more types of inputs, including instructions from another computer, which may be halfway around the world or out in space. For two computers to communicate with each other, a *modem* is required.

The methods of input vary a lot, but the result is always the same: The machine (hardware) receives information that it will store, process, and finally make public.

Memory A computer may perform thousands or millions of calculations in solving a problem. It must remember each piece of information it is given and keep track of all the processing it does en route to its final result. Also, each number, letter, symbol, and space put into a computer is turned into a group of eight bits (for example, the letter Z is translated as 01011010). Each bit must be stored somewhere and must be ready to be retrieved at a moment's notice (*see* RETRIEVAL). Almost constantly, then, a computer is storing something in memory and retrieving (accessing) something from memory.

Computer memory is astonishingly brilliant: it can store information swiftly and compactly; a 1000-page book can be stored on a rapidly spinning 8-inch (203-millimetre) disk in five seconds. It is also astonishingly stupid, without the faintest idea of what it is memorizing. It can store ones and zeros – nothing more – in the form of magnetized strips (which signify 1) or oppositely magnetized strips (which signify 0). Some types of storage use unmagnetized strips for zeros. Storage may also be in groups of electric switches in closed position (signifying 1) or open position (signifying 0). Such groups are called *flip-flop circuits* – flip and flop, 1 and 0.

A computer's memory is actually two memories: one permanent, one temporary.

ROM (read-only memory) ROM is permanent. Neither you nor the computer has any control over the information in ROM. Here the manufacturer has placed special data that can be used but not erased or changed. For example, when pressed, the pi key on a calculator releases a permanently stored number, 3.1415927, into the working circuits.

RAM (random-access memory) RAM handles all input, each electronic switch storing one bit. The more RAM a computer has, the more information it can store. However, RAM presents a special problem: it works only as long as an electric current flows through it. Without a current (the switch turned to 'off', the plug pulled out, a dead battery), all the stored information is wiped out.

Mass storage There is a way to get the best of both worlds — the power of RAM and the permanence of ROM. Suppose you're writing a novel. You wrote part of a chapter today, to be continued ... tomorrow? next week? Meanwhile, must you leave the computer turned on? What if somebody accidentally pulls the plug out? The solution is mass storage.

Information can be stored permanently on magnetic tape (much the way music is recorded on a cassette), on a FLOPPY DISK, a thin flexible disk coated with magnetic particles, or on a hard disk. These media store bits (zeros and ones) as magnetic strips running in either one direction (signifying 0) or the other (signifying 1). Mass storage devices called tape drives or DISK DRIVES can then retrieve ('read') the data, change it, and again store ('write') it onto tape or disk. Disk storage is much faster than tape, since the information is stored and retrieved randomly — that is, in any order — rather than having to be searched through from beginning to end. Much more information can be stored on tapes or disks than can be stored at any one time in the computer itself.

Memories, ah, memories ... Now we know what is stored (zeros and ones) and where it is stored (in magnetized strips or in switches), but what for? We don't just sit around and recollect them; we process them, in the CENTRAL PROCESSING UNIT (CPU).

When the input has been converted to zeros and ones, the computer's CPU takes over. The CPU consists of hundreds of computer chips, some handling the calculations or processing, others serving as the computer's memory. (In personal computers, a single chip, the *microprocessor,* takes care of the processing, and separate chips hold the memory.)

A piece of a CPU chip less than 6.5 millimetres (¼ inch) square can be made up of more than 100,000 tiny electronic switches lined up in rows and columns. Each switch is a storage unit, storing a single bit (a zero or a one). When a bit is a one, the computer turns a switch *on,* and a tiny electric current flows. When a bit is not a one (i.e., zero), the switch remains *off,* and no current flows. When information is processed, an electric current flows through various circuits, according to which switches are open or closed.

For example, a computer chess game contains a set of rules in switch form (a pawn can move this way, a knight can move this way *and* that, etc.). These rules are modified by the human player's input, based on intelligence, hunches, whims, experience, desperation, and so on, while the opposing player (the computer itself) modifies the rules by a set of grim logical steps called BOOLEAN ALGEBRA. Those steps are the ground rules of all computer processing, not just chess playing.

Output Now, after this once-over-lightly discourse on input, memory, and processing, we come to the output: the goal, the light at the end of the tunnel. The price of bananas, on a scale equipped with an LCD or LED display, is an output in its simplest form. Output can also appear on a CATHODE RAY TUBE — a TV or similar type of screen — or as a printout on paper (hard copy) produced by a PRINTER. For the sight-impaired, the printout can be rendered in braille.

A much more complicated output is the automatic pilot's electrical instructions to the engine throttles, rudder, elevator, and other flight controls. An example of in-between complexity is what happens in a word processor. Input: tap-tap by the genius at the keyboard. Processing: CPU jiggles the words into line length, checking spelling, hyphenating end words according to built-in memory, disgorging the output on a cathode ray tube, and then, if it meets the writer's approval, typing it on a printer.

See also RAM AND ROM.

COMPUTER GRAPHICS

Charts, graphs, diagrams, and pictorial images of all kinds displayed on a computer screen. These images can be generated in several ways — for example, from a memory bank: 'At the upper left corner, make a dot. On the second line, make a dot at the first space from the left. On the third line, make a dot at the second space from the left.', and so on. Such a set of instructions (called *point plotting*) would generate a slanting line. Any line or area can be point-plotted by inserting the proper directions (commands) into the computer memory.

Another method is to feed in mathematical data, such as '$C = \pi D$; solve for $D = 3$'. The result: a graphic display of a 3-centimetre circle.

Special graphics for various subjects can be stored in a computer and retrieved at the touch of a button. For example, chemical apparatus — test tubes, funnels, flasks, tubing, and the like — can be stored, retrieved, displayed, and printed this way. Architects, engineers, mathematicians, and other professionals have their special graphics available.

Graphics can be done by hand, too. A *light pen* touched to the face of a cathode ray tube (you know it better as a TV screen) draws or erases lines on the screen.

COMPUTERIZED AXIAL TOMOGRAPHY (CAT OR CT)

A method of obtaining a three-dimensional view of the interior of an object by building up a series of sectional views (Greek *tomo,* cut, section). This method is an elaboration of X-ray techniques, so let's take a brief look at them first.

Like light waves, X-rays are electromagnetic radiations, but they are much shorter than light waves and invisible to the human eye. X-rays can penetrate many objects, passing easily through soft substances such as fat and skin, less easily through denser tissue such as muscle, and hardly at all through bone and metal. That's how you can distinguish, in this illustration of an X-ray photo of someone's hand and wrist, the finger bones, the skin and muscle around them, and the metal ring. X-ray photography is enormously useful, but it is limited in two important ways:

1. To impress a useful image onto a photographic film, X-rays have to be fairly strong and are therefore possibly dangerous.

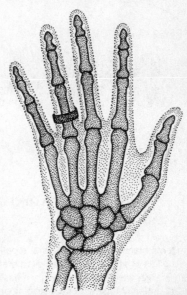

2. X-ray photography produces a two-dimensional image: an egg-shaped tumour might cast its true oval image from one angle but might show a circular image from another angle. (How many ways can you slice a hard-boiled egg?) Why not, then, make many images from many angles? There is a limit to the number of strong X-ray exposures that living tissue can safely tolerate.

We come, then, to the vastly superior (and much costlier) system called *computerized axial tomography* (CAT or CT). The system is sensitive to very small differences in density, so a doctor can, for example, examine a living brain or determine the exact size, shape, and location of a tumour or a blood clot. Another advantage of CAT is that there is no photographic image and therefore no need for strong X-ray beams. The image is constructed serially and sectionally by a computer. To see what that means, we can do a very simple analogical experiment using an electric torch and a light meter.

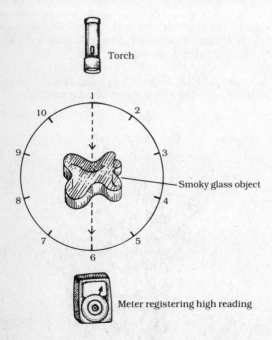

Here's the torch placed at the rim of a dial at point 1. At the centre of the dial is a piece of smoky glass. The light meter is at point 6. The torch's beam is weakened somewhat as it shines through the glass and strikes the meter. We get a certain reading and record it. Then

COMPUTERIZED AXIAL TOMOGRAPHY

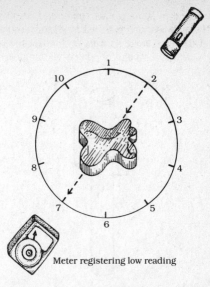

we move the meter to point 7 and the torch to point 2. This time the beam must pass through a greater amount of glass, so the beam is even weaker, giving a lower reading on the meter. Point by point, we move around the circle, getting a series of readings. These are fed into a computer, which assembles the numbers into a diagram of the glass. This diagram is displayed on a cathode ray tube (CRT), which resembles a TV picture tube.

The picture on the CRT is of just one slice (one *tomo*) through the glass. Suppose you were trying to construct the diagram of a long glass rod, round in some places and square in others. You could measure the thickness of the rod with the torch and meter, move the rod a short distance along its axis, and repeat the measurements, over and over. You would be doing a primitive kind of axial tomography.

Now for the real thing — the CAT scan. In place of a simple rod, we have a complicated human being. In place of a torch, we have a source of weak X-rays. In place of a light meter, there is a detector (a photomultiplier tube) that receives the X-rays and amplifies the reading. The readings are taken not only at pairs of points (1 and 6, 2 and 7, etc.) but by continual scanning around a whole circle.

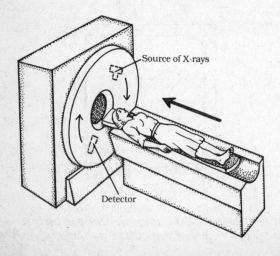

A movable stretcherlike table, which a technician operates by remote control, moves the patient slowly into the machine that houses the X-ray source and the detector. One scan is done, and the patient is moved forward a fraction of an inch. The scanning process is repeated, over and over, until the entire brain, liver, or other part of the body has been scanned and the readings fed to a computer.

Depending on the size of the organ(s) to be studied, and the detail desired — that is, how closely spaced the slices are to be — a CAT scan can take from a second or two to more than an hour.

The computer can be made to display the image of section after section of the body part on a cathode ray tube; as many sections as needed can be photographed for a permanent record. This drawing was copied from an image made at the level of the kidneys.

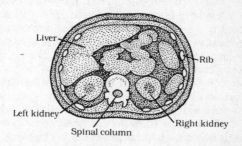

An offshoot of CAT is CIT, *computerized industrial tomography*. This is a system for inspecting machine parts, metal castings, and so on, for internal flaws. Metals resist the passage of X-rays more strongly than body tissues do, so a more powerful beam is required. Especially large or dense objects are examined with even more powerful GAMMA RAYS emitted by radioactive isotopes of caesium or cobalt. With either kind of ray, the human operators of the CIT machine must be carefully shielded from stray radiation.

COMPUTERIZED INDUSTRIAL TOMOGRAPHY
See COMPUTERIZED AXIAL TOMOGRAPHY

COMPUTER LANGUAGES

Computers are astonishingly fast and accurate; they can perform millions of precise calculations in a few seconds. But they suffer from one handicap: They have no idea what they're doing. They must be told what to do — a few steps at a time on a little pocket calculator, hundreds of sequential steps on a home computer, many times that on the big machines in offices and factories.

We give instructions to computers (that is, we *program*) through various languages devised for their suitability for different uses and crafts, trades, and professions. The languages consist primarily of short commands – PRINT, GO TO, DO, STOP – typed onto the computer's keyboard, also called a keypad. Each computer language has its own rules (syntax) and vocabulary that make one language more useful to an engineer, another to an accountant, and so on.

One of the simplest and most useful languages for a novice to learn is BASIC (*B*eginner's *A*ll-purpose *S*ymbolic *I*nstruction *C*ode), since it is built into almost all personal computers and can handle a wide variety of applications. Some other computer languages are COBOL (*Co*mmon *B*usiness-oriented *L*anguage), for business uses, such as general accounting; FORTRAN (*For*mula *Tran*slation) and ALGOL (*Algo*rithmic *L*anguage), used to help scientists with numerical problems; and PASCAL (named after the French mathematician and philosopher Blaise Pascal, 1623–1662), a newer general-purpose language often taught in the classroom because it is said to promote logical thinking. Surprisingly, computers can't 'understand' any computer languages. These languages exist for humans, as a convenient way of giving orders to the machine (hardware). As an analogy, consider a lift's push-button keypad (UP, DOWN, CLOSE, OPEN, 1, 2, 3, etc.). Those buttons 'speak' a language comprehensible to humans; computer people call it a *high-level language*. But a lift mechanic can work the machinery directly by turning this lever, closing that switch, and so on. His actions are directly involved with the machinery (*machine language*, a *low-level language*).

See also ASCII.

COMSAT *See* SPACE TRAVEL

CONDENSED-MATTER PHYSICS *See* PHYSICS

CONSTELLATION

A pattern of stars as seen from the earth. The stars, like everything else you see in the sky, are in motion. It's easy to observe that a relatively nearby object – the planet Venus, for example – changes its position from week to week against the background of stars. But stars are so distant (thousands to millions of times farther away than planets) that their motion relative to one another is imperceptible. When you pick out a pattern of stars in the night sky, the Plough, say, you are seeing almost the same pattern that people saw a few thousand years ago. They named the patterns after fancied resemblances to gods, heroes, or animals (think of Hercules, Orion, Leo the Lion, among many others), and myths were invented to explain how they came to be in the sky.

STARS OF THE PLOUGH AS SEEN FROM THE EARTH

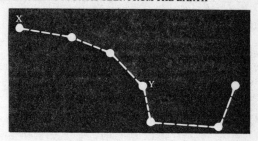

The particular patterns we see exist only from our viewpoint on the earth. For example, the stars of the Plough look like seven points of light on a screen. But no two-dimensional screen could accommodate the Plough, because each of its stars is at a different distance from us. Star X, for example, is more than three times as far away as star Y. Move a light-year or two to one side of the earth, and the seven stars form some other pattern.

**THE SAME STARS, FROM A VIEWPOINT
A FEW LIGHT-YEARS FROM THE EARTH**

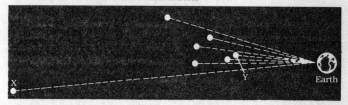

For convenience in locating stars, galaxies, and other celestial objects, modern astronomers have divided the sky into 88 areas; these are still called constellations, and many of them retain the classical names and parts of the old configurations. How else could you read about the latest astrophysical discovery in a constellation that is really a princess who was changed into a bear? All right — the princess was Callisto, and the bear (and constellation) is Ursa Major, the Great Bear.

CONTINENTAL AIR MASSES, POLAR AND TROPICAL See AIR MASS ANALYSIS

CONTINENTAL DRIFT See PLATE TECTONICS

CONTROL, SCIENTIFIC See SCIENTIFIC TERMS

CONTROL RODS See NUCLEAR REACTOR

COOLEY'S ANAEMIA

A hereditary blood disorder, also known as *thalassaemia*.
See also GENETIC DISEASES.

COPOLYMER 1 (COP 1) See NEUROMUSCULAR DISORDERS

CORIOLIS EFFECT

Also called the *Coriolis force,* it was first described by the French physicist Gaspard de Coriolis (1792–1843). A free-flowing substance (e.g., wind, water current), moving in a northerly or southerly direction, will be deflected towards the right if north of the equator, to the left if south of the equator. This is a matter of considerable importance to meteorologists, navigators, aeroplane pilots, launchers of space vehicles, artillerymen, and a certain pair of cockroaches.

Cockroach A, intelligent, evil, and armed, stands at the centre of a disc that is rotating counterclockwise, from west to east (as the earth does when viewed from above the North Pole). Cockroach A aims at his enemy, cockroach B, and fires. His tiny bullet flies in a straight line, but during flight its target, B, is carried by the disc's rotation to a new position, B_1. So on this disc, rotating counterclockwise, a shot from the centre, aimed towards the rim, seems to veer off to the shooter's right (and misses its target).

Act 2 of the cockroach vendetta: B, unscathed, aims at A and

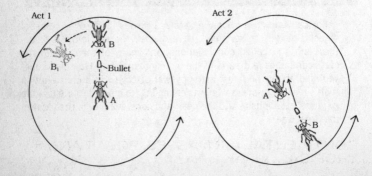

fires. But while he was aiming and firing, he and his gun were being carried sideways towards the right in relation to A. The bullet too was already moving sideways to the right as it emerged from the gun. So it passed to the right of Cockroach A.

Act 3: Both evildoers start dashing hither and yon, aiming and firing at each other. The bullets, when properly aimed (as seen in the gunsight) fly to the right of their target.

Act 4: The cockroaches have now dashed around to the underside of the disc (cockroaches, as you well know, can walk on the underside of a surface, and Australians don't even regard themselves as being under anything — because they're not). Same rotation of the disc (or globe, from our expanded scope), west to east, but here this is a clockwise motion, and bullets fly to the left of their targets. See for yourself: look at the diagrams through the other side of the page.

South of the equator to the left, north to the right — the Coriolis effect operates on all freely moving objects: the Gulf Stream and other currents, rockets after launching, great winds and little zephyrs, even (to a tiny extent) footballs. Its effect is greatest on north-south motion and least (in fact, zero) on east-west.

CORONA *See* SOLAR SYSTEM

CORONAGRAPH *See* SOLAR SYSTEM

CORONARY ARTERIES *See* HEART ATTACK

CORONARY ARTERY OCCLUSION
See HEART ATTACK

CORONARY BYPASS SURGERY *See* OPEN-HEART SURGERY

CORTISONE *See* STEROIDS

COSMIC RAYS

Atomic particles composed mainly of hydrogen nuclei. They are believed to originate in the sun and in explosions of old stars. Primary cosmic rays streak through space at nearly the speed of light, giving them tremendous energy. Nearing the earth, they collide with molecules of the atmosphere, shattering these and producing the far less energetic secondary cosmic rays, which can reach the earth's surface; fortunately for us, very few of the primary rays do so (*see* RELATIVITY.)

Even the secondary rays have enough energy to penetrate the

nuclei of plant and animal cells and can cause mutations (hereditary changes) in them by altering their genes.

COSMOLOGY

In the beginning there was (tick one):

- The void
- A huge turtle on whose back the earth rested
- A group of crystal spheres-within-spheres surrounding the earth, in which were imbedded the sun, moon, and stars
- The big bang
- Another theory
- No choice

Your choice, or lack of choice, depends on which aspect of cosmology — the study of the origin, composition, and dynamics of the universe — you favour. All the choices listed are current in folklore or religion or science or personal hunch. The overwhelming choice of scientists is the big bang, which on the face of it seems utterly preposterous. According to the big bang theory, the universe began about 12 to 15 thousand million years ago as a tiny sphere the size of, what, a basketball? (a tennis ball? a pinhead? smaller?) containing *all* the matter and energy that then became the universe. This tiny, unimaginably dense, impossibly heavy object exploded and expanded. Before we proceed with the details of such an unbelievable occurrence, let's see on what evidence scientists base their belief.

Tomes have been filled with data of observations and calculations supporting the big bang theory. There are two major streams of supporting data, which can be grossly classed and nicknamed the 'raisin cake' and the 'fire in the cave'. More elegantly, they are dubbed the expanding universe and background radiation.

The expanding universe Imagine an uncooked raisin cake, made of uncooked cake mixture (naturally), with raisins scattered evenly throughout — inside and on the surface. The cake is put into a hot oven and begins to bake. It puffs up, expanding in all directions. The distance between any raisin and all the other raisins increases. A raisin gifted with vision would see all the other raisins receding from it. Gifted with intelligence as well, such a raisin would conclude that the cake is expanding.

In 1929 the American astronomer Edwin Hubble (1889–1953) presented evidence that the universe is expanding (see DOPPLER EFFECT). The thousands of millions of galaxies, each containing thousands of millions of stars, are all receding from one another at tremendous speeds. The universe is expanding.

From what beginning? Run the film projector backwards and we come to some sort of starting point, a time when all the objects of the universe come together, condense into — according to most cosmologists — the tiny massive sphere that underwent the big bang.

Background radiation Imagine entering a cave and finding a heap of warm ashes. You also notice that the air is warm. You conclude that there has been a fire, the cause of it all.

The 'warm ashes' of the universe are radioactive substances whose rate of 'cooling down' (radioactive decay) points to a starting time. The 'warm air' is the presumed empty space of the universe, which was formerly thought to have a temperature of ABSOLUTE ZERO, the temperature of nothing. But two American physicists, Robert Wilson and Arno Penzias, in 1965 announced their findings: space does have a temperature, 2.7°C above absolute zero (i.e., a temperature of 2.7 K or −270.5°C). 'Empty' space, it turns out, is swarming with energetic, heat-inducing photons that were liberated shortly (about 700,000 years) after the big bang. That event fits neatly into the timetable of the total big bang history, from the first billionth of a second to the present day — and into the future.

What about the future? No gypsy fortune-teller ever offered such bizarre possibilities: 'Some say the world will end in fire. Some say in ice'.

At the moment of the big bang, the temperature of the tiny universe was over 100,000 million degrees Celsius. Immediately, with explosion and expansion, there was a drop in temperature, thousands of millions of degrees in a fraction of a second. With cooling came a drop in the pressure that originally blew the primordial thing apart. But the momentum is still with us; the pieces of the universe are still flying apart. However, another force was operating during the big bang, and it is still in action; GRAVITATION, pulling every particle of matter in the universe to every other particle, is working against momentum. Which will win? (Talk about cliff-hangers — you will only have to wait several thousand million years.) If momentum prevails, the universe will continue to expand and cool, its available energy diffused into an icy death (see ENTROPY). Such a fate is blandly defined as the '*open universe*' and is regarded by most cosmologists as the likely end.

A less gloomy prognosis assumes that there is sufficient matter in the universe for gravitation to prevail, in which case pulling together will overcome flying apart, so that the momentum will be slowed, brought to a stop, and reversed. The film projector will run backwards, back to the original primordial sphere, incredibly hot, unimaginably compressed and heavy.

And then? If the tug-of-war figures are right, another big bang, another cycle of expansion, another era of formation into atoms and molecules, and then dust clouds and stars and GALAXIES — another phase of the OSCILLATING UNIVERSE.

COWLING

The streamlined cover of an aeroplane engine.
See also AERODYNAMICS.

COWPOX   See IMMUNE SYSTEM

CPU   See CENTRAL PROCESSING UNIT

CRAB NEBULA   See STELLAR EVOLUTION

CREATIONISM   See EVOLUTION

CREATION SCIENCE   See EVOLUTION

CRITICAL MASS   See NUCLEAR REACTOR

CROSSOVER NETWORK   See LOUDSPEAKER

CROSS TALK

The unwanted transfer of information from one communication system or channel to another. A common example is the nuisance of a second conversation faintly heard on a telephone. This may be due to mixed-up connections or to a phenomenon called *inductance,* in which a strong electric current induces a weak current in a nearby wire. By extension, the term refers to any such interference, as for example in a computer-to-computer dialogue.

CRT   See CATHODE RAY TUBE

CRUST OF THE EARTH   See ROCK CLASSIFICATION

CRYSTALLOGRAPHY   See PHYSICS

CT SCAN   See COMPUTERIZED AXIAL TOMOGRAPHY

CVS   See AMNIOCENTESIS

CYCLAMATES See SWEETENING AGENTS

CYCLE (IN ELECTRIC CURRENT)
See DC AND AC; RADIO

CYCLOSPORINE See IMMUNE SYSTEM

CYCLOTRON See PARTICLE ACCELERATOR

CYSTIC FIBROSIS

A hereditary disease in which major glands malfunction, with eventual involvement of the lungs.
See also GENETIC DISEASES.

CYTOLOGY See BIOLOGY

CYTOSINE See DNA AND RNA

DAISY WHEEL PRINTER *See* PRINTER

DC AND AC

Abbreviations of *d*irect current and *a*lternating current. The big batteries in cars and the little ones in electric torches are essentially tanks of electrons. Batteries deliver electrons in a steady stream — direct current — until exhausted or recharged with more electrons.

Alternating current is a stream of electrons that alternates — changes direction — at a regular rate. The rate in Britain is 50 hertz (abbreviated to Hz), meaning 50 back-and-forths, or *cycles*, per second. Alternating current is produced by AC generators (*see* GENERATORS AND MOTORS, ELECTRIC).

Alternating current has certain advantages over direct current, the principal one being that it can be sent through a TRANSFORMER, which quietly and efficiently changes the voltage to whatever is currently (pun) desired. For example, a doorbell runs on about 6 volts; a TV picture tube needs 20,000 volts or more (notice the 'DO NOT OPEN' warning on the back of your TV). Yet both can operate off the 240-volt house current, thanks to transformers.

DECAY *See* BIODEGRADABILITY

DECIBEL

One tenth of a bel, a unit of power named in honour of Alexander Graham Bell (1847–1922), inventor of the telephone. Decibels (db or dB) are commonly used as measurements of sound power or loudness. Doubling the loudness or power of a sound adds 3 db to its rating. Thus a sound of 21 db is twice as loud as a sound of 18 db. Some common decibel ratings:

Whisper	15– 30 db
Clock ticking	20– 40 db
Conversation	30– 60 db
Discotheque	105– 115 db
Nearby thunder	120– 130 db
Explosion	120– 140 db

DECIMAL SYSTEM *See* BINARY NUMBER SYSTEM

DECLINATION See CELESTIAL COORDINATES

DEDUCTION See SCIENTIFIC TERMS

DEGENERATIVE JOINT DISEASE See ARTHRITIS

DEGRADABLE See BIODEGRADABILITY

DEOXYRIBONUCLEIC ACID See DNA AND RNA

DEOXYRIBOSE See DNA AND RNA

DESERTIFICATION

The process of becoming a desert (Latin *deserere*, to abandon). A desert is a region in which the vegetation is so scanty as to be incapable of supporting any considerable human population. Deserts are bad enough news in a world whose human population is growing dramatically; much worse is the fact that the deserts themselves are growing.

Experts at the United Nations estimate that 10% of the earth's people have already been affected to some extent by desertification. In North and Central Africa, the most severely affected regions, famine has killed hundreds of thousands of people in recent years.

A long-standing cause of desertification has been prolonged drought, although today the main cause is probably large-scale human activity. To feed the rapidly growing number of people, infertile land is overused to the point where little or nothing will grow on it — a desert in the making. Overgrazing by increased numbers of cattle and indiscriminate cutting of trees for timber and firewood lead to soil erosion and more desertification.

Some experts believe that the process may be evolving into a cycle: Areas without vegetation reflect more sunlight, beginning a chain of events that ends with an increase in the amount of dry air at ground level. The dry air promotes the growth of the desert, which in turn reflects more light, and so on, ominously.

DEUTERIUM See NUCLEAR ENERGY

DIABETES

A group of disorders in which the body uses carbohydrates (sugars and starches) and fats in an abnormal way. Diabetes is a complex, incurable disease with various suspected causes: heredity, infections, pancreatic disease, obesity, drugs, pregnancy, and combinations of these and others. It is estimated that up to one million people in the United Kingdom have diabetes.

Before we consider the disease, let's look at the normal pattern, much simplified.

The cells of the body use energy constantly. The source of the energy is *glucose*, a basic simple sugar obtained from food. The glucose circulates in the blood and reaches all the cells.

The *islets of Langerhans*, clumps of special cells in the pancreas, react to changes in the level of glucose in the blood. If the level rises, as when digested food enters the blood, one group of these, the *beta cells*, secrete *insulin*. This HORMONE helps the body cells in absorbing glucose. It also helps to convert glucose to *glycogen*, a starchlike substance that is stored mainly in the liver.

A second set of islet cells, the *alpha cells*, work in the opposite way. If the glucose level drops, they secrete *glucagon*, which promotes the conversion of the stored glycogen to glucose, bringing the level up. (These *glu-* and *gly-* words are based on the Greek root for 'sweet'.)

In effect, then, insulin and glucagon operate as a checks-and-balances team to keep the glucose level exactly right. The balance is upset in diabetics because too little insulin, or none, is secreted or because the body cells are unable to use it. As a result, an excess of glucose circulates in the blood. Some of the excess is filtered out by the kidneys and appears in the urine.

Two main kinds of diabetes exist: *juvenile onset* and *maturity onset*. The juvenile type (also called insulin dependent) usually appears before the age of 20 and is characterized by a severe shortage or complete lack of insulin. Unable to derive sufficient energy from glucose, the body begins to draw on stored fat. The chemical breakdown of the fats results in an excess of substances called *ketones*, leading to a dangerous condition called *diabetic acidosis*. If injections of insulin are not provided promptly, the victim goes into *diabetic coma*, followed by death. A special problem with insulin injection is the danger of overdosing enough to cause *insulin shock*. Prompt treatment with sugar is needed to restore the balance.

Maturity-onset diabetes (also called non-insulin dependent) is a much less serious disorder. It appears mainly after the age of 40 and accounts for about 80% of diabetic cases. Insulin is secreted but is not taken in by the body cells. Low-carbohydrate diet, exercise, and weight control are sufficient to control the disease in most patients, and drugs are available to enhance the response of the body cells to insulin.

In severe diabetes the person loses weight, weakens, and may suffer damage to blood vessels and nerves. Long-term complications of the disease include kidney problems, heart disease, stroke, and blindness.

DIALYSIS OF BLOOD   See KIDNEY DIALYSIS

DIASTOLIC PRESSURE   See BLOOD PRESSURE

DIE PRESSES   See MANUFACTURING PROCESSES

DIES   See MANUFACTURING PROCESSES

DIFFERENTIAL CALCULUS   See CALCULUS

DIGITAL AND ANALOGUE

Two methods of indicating a quantity or measuring a value. Digital (Latin *digitus*, finger) refers to counting or measuring by distinct units. When you count on your fingers, any finger is a distinct unit, equal to each of the other nine fingers. A digital watch, similarly, counts in distinct units — seconds — adding them up into minutes and hours and displaying these units as actual numbers, for example, 12:43.16.

Some other digital instruments include a car odometer, which counts miles and tenths of a mile, and a digital-type electronic thermometer, which displays temperature in degrees and tenths of a degree.

Analogue can be roughly defined as 'bearing a relationship to'. You say, 'I'll look at the time', but if you are using the older type of watch face, what you see is not the time but an *analogue* of the time — the continuously varying angle between the hour hand and an imaginary vertical line and a second continuously varying angle between the minute hand and the vertical line. You long ago learned to measure

these angles so accurately that you don't have to look at the numbers; in fact, some analogue watches don't even have numbers.

Some other analogue instruments are liquid-containing thermometers, in which the length of the column of liquid varies with the temperature; speedometers, on which the angle of the pointer varies with the speed; and bathroom scales, on which the angle of the pointer varies with the user's weight.

DIGITAL RECORDING See SOUND RECORDING

DIODE See LCD AND LED

DIOXINS

A group of chemical compounds, one of which, called 2, 3, 7, 8-TCDD (*tetrachlorodibenzo dioxin*), ranks among the most toxic substances known. It is found in small amounts as a contaminant in weed killers (herbicides) and is extremely long-lasting. Animal experiments have shown that extremely small doses cause liver and kidney damage, birth defects, and damage to the immune system. Dioxins are also thought to be carcinogenic.

During the Vietnam War the United States sprayed a herbicide called *Agent Orange* over millions of acres in Vietnam to defoliate areas in which enemy troops might take cover. Thousands of soldiers on both sides who were exposed to the dioxin developed extensive skin rashes. Many studies have been done to determine whether other health effects can be traced to the exposure. So far, the results are not conclusive.

DIRECT CURRENT See DC AND AC

DIRECT CURRENT, GENERATION OF
See GENERATORS AND MOTORS, ELECTRIC

DIRIGIBLE See BALLOON

DISCOMFORT INDEX See TEMPERATURE-HUMIDITY INDEX

DISK, COMPUTER See FLOPPY DISK

DISK DRIVE

A device used for driving disks that store computer data. It resembles your record turntable in two functions: (1) it spins a disk, from which it (2) picks up (or 'reads') data and feeds that data into a

COMPUTER. It can also feed (or 'write') data onto a FLOPPY DISK to be stored for later use. Like the magnetic head in a tape recorder, which allows you to record and play back music, the read/write head in a disk drive places data (in the form of BITS) magnetically on a floppy disk as it spins.

Computer disk drives cost more than most record turntables because more is required of them. On a record turntable, it's enough to set the stylus at the beginning of the record, or wherever you please. The stylus and pickup will keep going from that point, guided by the spiral groove that contains the musical or spoken data. A computer disk has no grooves, and it doesn't store sequential data that you want to play from beginning to end. Instead, the information is placed on the disk randomly, with one section of the disk serving as a kind of table of contents, telling the disk drive where to find specific data. The disk doesn't have to be played back from beginning to end to find the right data.

DISKETTE   See FLOPPY DISK

DISSEMINATED SCLEROSIS
See NEUROMUSCULAR DISORDERS

DISTILLATION   See ANALYSIS

DITIOCARB   See AIDS

DNA AND RNA

Abbreviations of *d*eoxyribo*n*ucleic *a*cid and *r*ibo*n*ucleic *a*cid. These two classes of compounds are *nucleic acids*. They are the agents for the design and assembly of proteins, which are giant, complex molecules, the basic material of all life.

First, a brief overview of the crucial services performed by various proteins:

- They are the structural materials — the girders, concrete, and timber — of all living cells.
- They participate in thousands of chemical and physical reactions that spell life. Some of these reactions are in turn facilitated by other proteins called ENZYMES and are controlled by still other proteins called HORMONES.
- They protect us, in the form of disease-fighting antibodies produced by the IMMUNE SYSTEM.

Even though there are thousands of proteins, they are all assembled in the same manner, by means of the master blueprints and instruction manuals called DNA and RNA. Some proteins are incorporated into skin cells, others into muscle cells, nerve cells, bone cells, and so on. Yet inside each cell, tucked away in the nucleus, is an exact copy of the owner's original moment-of-conception DNA. (We'll deal with RNA later.) DNA resembles a spiral staircase — the well-known double helix, seen here in a short section of a model (A). The structure is easier to understand in this simplified illustration (B), where the untwisted DNA resembles a ladder. Each side of the ladder is a long chain of molecules, *phosphates*, alternating with a type of sugar called *deoxyribose* (C). Attached to each sugar molecule is another molecule, like part of a ladder rung, called a *base*. A single group — base, sugar, phosphate — is called a *nucleotide* (D). A very small package of life, such as a virus, has over 5000 nucleotides. A single human cell, engaging in a far more complicated lifestyle, contains, as you might expect, a far greater number of nucleotides — over 5000 million! Nevertheless, whether it's in you, in a virus, or in a geranium, each nucleotide consists of a base-sugar-phosphate trio.

There are four kinds of bases: *guanine, cytosine, adenine*, and *thymine* (abbreviated G, C, A, and T). The bases (partial rungs) join the sides of the ladder together, making full rungs. A full rung consists of two bases and is therefore called a *base pair*. There are different sizes of bases, and they have different chemical natures. This limits the kinds of possible pairs to only two, T with A, C with G. Yet the almost countless proteins that form the tissues of life — bone, nerve, blood, skin, cartilage, petal, pollen, fish scale, hair — all are assembled at the direction of strands of DNA, consisting of monotonously repeating phosphate and sugar molecules joined across by pairs (only two kinds!) of bases. How can such enormous variety arise out of such simplicity? (And how can you possibly stop reading now? Read on.)

A human embryo begins with fertilization, the union of a sperm cell and an egg cell. (This is, of course, true of many other forms of life, but let's stick to our own for now.) Each sperm cell and each egg cell brings a dowry — its DNA — coiled up and divided among 23 rod-shaped bodies called *chromosomes*. The contents of the 23 pairs (from sperm and egg) are the embryo's *genes* (estimated number between 50,000 and 100,000) — its entire biological heritage. The embryo develops by dividing into two smaller cells; these grow, divide into four cells, then eight, sixteen, and so on. Each of these 'daughter' cells must have a full complement of DNA, identical to the DNA in the original one-celled embryo (we'll see why shortly). This need is satisfied by a process called *replication*.

Replication (doubling) The nucleus of a cell holds the DNA,

DNA AND RNA

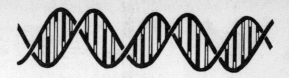

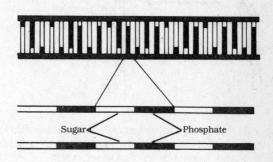

ONE NUCLEOTIDE

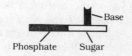

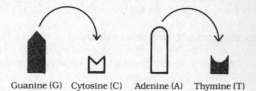

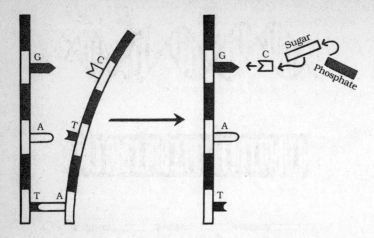

along with raw materials for making more DNA and enzymes to speed the replicating (doubling) process. At the start of replication the chemical bond between each pair of bases loosens, allowing the left and right sides of the chain to separate in ziplike fashion. The newly uncovered end of each base can then bond to another molecule that fits it.

The G on the left half of the chain bonds to an available C, for example; the newly bonded C bonds to a sugar molecule, which in turn bonds to a phosphate molecule, and so on, up and down the left side of the chain; meanwhile, similar events are taking place on the right side. When it's all over, the original DNA chain has become two identical chains.

When the cell divides, each daughter cell gets one of the chains. And when a daughter cell divides, the replication process will provide copies of DNA for the subsequent divisions.

Transcription (copying the blueprint) The 'protein factory' is the *cytoplasm*, the living material that makes up most of the cell. But the DNA — the master blueprint — is separate, in the nucleus, where it remains a reference work to be consulted throughout the life of the cell. To carry protein-assembling instructions out into the 'factory', the DNA makes partial copies of itself, called *messenger RNA*.

Why is it called *R*NA? This molecule contains *ribose*, a sugar with one more atom of oxygen than the *deoxyribose* of DNA. There are some other differences as well: RNA is usually a single strand, unlike the double-stranded helix of DNA, and RNA contains *uracil* (U) in place of the closely related thymine (T) as one of its bases.

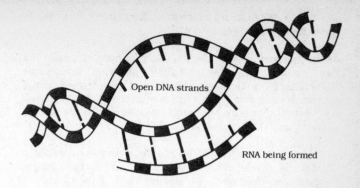

Why does the DNA make only partial copies? Every cell nucleus holds the entire DNA blueprint, but any given kind of cell needs only a small part of the blueprint. For example, muscle cells must make enzymes for extracting large amounts of energy, but they have no need of instructions for making enzymes that build bone, skin, or other kinds of tissue. How then do your cells avoid such needless duplication? Behold: as transcription begins, the strands of DNA open, but only in the sections where genes are to be copied. Other genes nearby (called *operators, regulators*, and *promoters*) act as on/off switches to limit the length of the copy to what is needed.

Raw materials in the nucleus bond to the newly uncovered bases on one strand of DNA. Thus is built a strand of messenger RNA, which then moves out of the nucleus into the cell cytoplasm, carrying the instructions for protein synthesis.

We now get back to the problem of how a mere four bases can determine the nature of the thousands of different giant, enormously complex protein molecules. Maybe the bases are a code for amino acids? Proteins are chains — very long ones, usually — of amino acids. There are 20 amino acids, and they bond to each other, head to tail, head to tail, in almost limitless numbers and sequences.

Now, are four bases enough to act as a code for 20 amino acids? No, but groups of bases are. A group of three bases forms a *triplet code* (ACC, GGG, CGU, etc.) with 64 possible combinations (you can check it out for yourself), more than enough for 20 amino acids. Thus the triplet ACC is the *codon* for a particular amino acid, histidine. UGU codes another amino acid, valine, and so on. That's how a strand of nucleic acid 1000 nucleotides in length can direct the synthesis of an average-sized protein composed of hundreds of units of amino acids.

In the complicated manoeuvres of genes, over and over, an occasional mistake caused by chemicals or radiation may occur. This results in a *mutation*, a change in a characteristic that may be passed to subsequent generations.

Assembling proteins Protein synthesis in the cytoplasm of a cell is carried out on large numbers of *ribosomes*, the granular dotlike objects in the drawing on the next page (copied from a photograph made through an electron microscope). The large dark object in the centre is freshly made protein. Here are the steps in the synthesis, much abbreviated and simplified:

1. An arriving messenger RNA molecule drapes itself around a ribosome, with its codons in position for 'reading' by molecules of *transfer RNA*, which are also produced in the nucleus. A segment of transfer RNA is very short, and it is specific for one amino acid. For example, it will pick up a free molecule of histidine, but no other amino acid, from the cytoplasm.

   The specificity, or 'choosiness' of transfer RNA molecules is itself based on a genetic code, which scientists had deciphered by 1988. That step offered the possibility that useful new proteins could be synthesized through GENETIC ENGINEERING.
2. At the proper codon on the messenger RNA, the transfer RNA releases its amino-acid burden. Similarly, other amino acids are deposited by their specific transfer RNA carriers, to be fitted into the ever-lengthening chain of protein. Each amino acid is joined to its neighbours by a *peptide bond*, a type of link unique to proteins.
3. A completed protein peels away from the messenger RNA, which is thus left free to repeat the process a limited number of times.

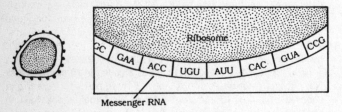

Look at one of your fingertips. Find a single ridge of one fingerprint. In a piece of that ridge about the size of the full stop at the end of this sentence, there are more than 2000 cells. Each single cell was constructed in a series of steps described by this entire article on DNA and RNA. Can you think of a science fiction story as fantastic?

See also PROTEINS AND LIFE.

DNA FINGERPRINTING

A technique, also called *genetic fingerprinting*, for identifying the component of DNA (the material of the genes) that is unique to a particular individual. Just as one person's fingerprints are different from everyone else's and can be used for identification, so a small section of the DNA of an organism (which is present in every cell of the body) uniquely distinguishes that particular organism from all others.

Most of an organism's genes go to making that organism what it is — person, cat, daisy, or what-have-you. Therefore, not surprisingly, the differences in genetic makeup between two individuals of the same type are very small. However they do exist. These varying bits of genetic material take the form of sequences of DNA, called *mini-satellites*, which are repeated several times. The number of repetitions of a mini-satellite per region of a gene can vary enormously between unrelated individuals. Chemical analysis of an organism's DNA from a sample of blood, tissue, semen, etc., using the techniques of chromatography and electrophoresis (see ANALYSIS), produces a two-dimensional pattern of spots. This corresponds to the genetic profile of that organism, complete with the repeating sequence which can be picked out.

Although comparatively new, DNA fingerprinting is becoming an established forensic technique and has been used successfully as evidence in court cases of rape in Britain. It is also being used to investigate family relationships in animal populations, and to measure the extent of inbreeding by looking at the degree of variability in DNA profiles. Comparing the DNA patterns of parents and offspring may in future be useful in preventing trade in endangered species or in proving paternity suits.

DNA LIGASE *See* GENETIC ENGINEERING

DOLBY NOISE-REDUCTION SYSTEM

Electrical circuitry that eliminates the annoying hiss that otherwise accompanies tape-recorded music or speech. The hiss is most evident during silent intervals or soft passages, with the volume control turned up high. It is an inevitable product of the tape-recording process itself.

A blank (unrecorded) tape has a coating of magnetic particles compactly but randomly impressed on the tape's surface. During the recording process, sound waves are converted to magnetic vibrations. The vibrations force the particles to line up in rows, one row for each vibration. The lining-up process involves a lot of jostling and bumping among the particles, and this is what produces tape hiss.

The Dolby noise-reduction system (named after its developer, Ray Milton Dolby, born 1933 – an American engineer) is ingenious and simple. During recording, a special circuit responds to the soft passages (A) by amplifying them (B). This happens *before* the sound signals reach the tape. The jostling of the magnetic particles produces the usual tape hiss, but now the strength of the desirable sound compared to the hiss is much greater (the signal-to-noise ratio is higher). Finally, when the tape is played (C), a deamplifying circuit reduces the amplified soft passages to their former volume and at the same time reduces the tape hiss in the same proportion. The result is a hiss inaudible, or almost inaudible, to human ears. The same thing happens when the sound track on a cinema film is subjected to Dolby treatment.

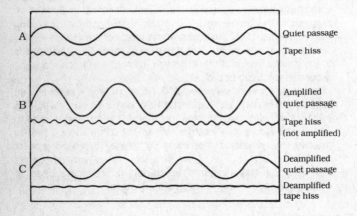

DOPAMINE See PARKINSON'S DISEASE

DOPING (IN TRANSISTORS) See SEMICONDUCTORS

DOPPLER EFFECT

Responsible for seemingly unrelated phenomena, such as the terrifying rise in pitch heard as a lorry hurtles towards you with horn blaring and the *red shift* in the light from a receding star. It is also the operating principle for various speed-measuring devices. We doff our hats to Christian Johann Doppler (1803–1853) for his explanation of the effect.

Regular repeating actions such as sound waves, light waves, or radio waves have a certain frequency, or number of waves per second. There is a change in the perceived frequency if the source of the waves (the lorry, for example) moves in relation to the receiver (you). As you and the lorry approach one another (either or both may be moving), there is an increase in frequency, which you perceive as a rise in pitch. When you recede from one another (again, either or both may be moving), there is a decrease in frequency, which you hear as a drop in pitch. When there is no relative motion (both standing still or both moving at the same speed in the same direction), there is no change in frequency.

What explains all this? Imagine yourself at one end of a swimming pool. At the other end is a machine that pats the water at a regular rate, say once per second, producing waves at a frequency of 60 per minute. Standing where you are, without motion relative to the source of the waves, you receive the same frequency, 60 waves per minute. Suppose you now start to swim towards the source; you receive more than 60 waves per minute, because you're picking up some extra waves by advancing towards the source. The frequency (the number of waves slapping against you) rises. Turn and swim away from the source, and the frequency drops below 60.

If you found a cooperative lorry driver to drive towards you, horn blowing constantly, at 30 miles per hour, then 40, 50, and 60, you'd find that the faster he drove, the greater would be the change in pitch. You could even, in this way, make a rough estimate of the truck's speed. As a matter of fact, there are devices that use the Doppler effect as the basis of speed-measuring systems. Let's look at a couple.

Police radar speed checks A police car equipped for catching speeding vehicles is parked alongside a road. It sends out radio waves that strike vehicles on the road and are bounced back to a receiver in the car. The receiver compares the frequency of the outgoing and reflected waves, converts the information to miles per hour, and displays the speed on a dial.

Suppose the waves hit a parked vehicle. The outgoing and reflected waves have the same frequency, because there is no relative motion between the police car and the other vehicle. If a vehicle is moving at moderate speed, there is a moderate difference in the frequencies of the waves. But a fast-moving vehicle produces a big difference. The speeder is flagged down, thanks to Herr Doppler's effect. Similar radar devices in airport control towers use the Doppler effect to determine the speed of aircraft in the area.

Astronomical speed checks Like sound waves or radio waves, light waves can be used to measure speed. An instrument called a *spectrometer* spreads light into a rainbowlike spectrum of thousands of parallel lines. The spectrum ranges from violet, with the greatest wave frequency, through blue, green, yellow, and orange to red, which has the lowest frequency.

Recall that the increase in frequency of the sound waves was detected as a rise in pitch when the lorry approached. The wave frequency of light from some luminous object — a star, for example — similarly increases if the star moves towards us. Coupled to a telescope, a spectrometer detects this movement as a shift of the spectral lines towards the blue side (the *blue shift*). A receding star would cause a shift towards the red side. The amount of the shift is a measure of the speed with which the star is approaching or receding from the earth.

Much of today's understanding of the universe is based on spectrometry. The theory of the expanding universe was confirmed by the discovery of the red shift of starlight from other galaxies, an indication that the distance between them and our galaxy is increasing.

See also COSMOLOGY.

DOT-MATRIX PRINTER   See PRINTER

DOUBLE-BLIND TEST   See SCIENTIFIC TERMS

DOUBLE STARS   See BINARY STAR

DOWN'S SYNDROME

A serious disorder, formerly also known as *mongolism*, that is congenital (present at birth), and occurs mainly in children born to older women. Its characteristics include upwards slanting eyes, a small head, various other physical abnormalities, mental retardation, and shortened life expectancy. The syndrome is caused by the presence of an extra chromosome in the body cells (the normal number is 46). Chromosomes are repositories of the material of heredity, deoxyribonucleic acid.

Down's syndrome can be diagnosed early in pregnancy under the microscope, by a count of the chromosomes in the fetal cells, obtained by a method called AMNIOCENTESIS.

See also DNA AND RNA

DRAG   See AERODYNAMICS

DRUGS, BRAND-NAME AND GENERIC
See GENERIC DRUGS

DURALUMIN

A light, strong aluminium ALLOY, widely used in aircraft.

DYKE (GEOLOGICAL) See ROCK CLASSIFICATION

EARTHQUAKE See PLATE TECTONICS

EBV See HERPESVIRUS DISEASES

ECG See ELECTROCARDIOGRAPH

ECHOCARDIOGRAPH See ULTRASONICS

ECHOLOCATION See SONAR

ECLIPSE

The passage of one astronomical body into the shadow of another. For example, the earth sometimes passes into the long, cone-shaped shadow the moon casts into space. The sun's light may be blocked out or occulted, causing a *solar eclipse*. Several conditions determine what kind of solar eclipse, if any, will be seen.

The moon comes between the earth and the sun every month (new moon), but for an eclipse to occur the sun, moon, and earth must be in a straight line. When an eclipse does occur, the moon's shadow sweeps across part of the earth; to people under the central part of the shadow (*umbra*), it appears that a small curved bite has been taken out of the sun's edge. The bite grows and grows until the entire sun is gone. Only the soft glow of the corona, the outermost part of the sun's atmosphere, is visible. This is a *total eclipse*. Totality lasts a few minutes at most. To people in the outer part of the shadow (the *penumbra*), the bitten-out edge appears, grows, shrinks, and disappears without reaching totality — a *partial eclipse*.

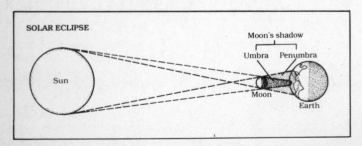

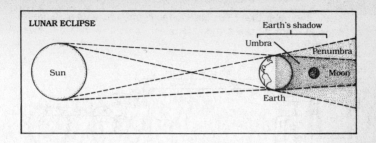

Because the moon's distance from the earth changes, its umbra sometimes falls short of the earth, leaving a ring of sunlight. This is an *annular eclipse* (Latin *annulus*, ring).

In a *lunar eclipse*, the earth comes between the sun and the moon, and the earth's shadow falls on the moon. The earth's shadow is wide and the moon is comparatively small, so a lunar eclipse can last for hours. The moon grows much dimmer and takes on a coppery hue, but it never actually disappears.

See also OCCULATION.

ECLIPSING BINARY STARS   *See* BINARY STAR

E. COLI   *See* GENETIC ENGINEERING

ECOLOGY   *See* BIOLOGY

ECT   *See* ELECTROCONVULSIVE THERAPY

EDISON PHONOGRAPH   *See* SOUND RECORDING

EEG   *See* ELECTROENCEPHALOGRAPH

ELECTRICAL ENERGY   *See* ENERGY

ELECTRICAL UNITS

Many of us spend more on electricity than on any other not-for-free source of energy, such as petrol. Yet most of us know relatively little about what we're buying. Herewith are offered some brief explanations of the various electrical units to be seen on toasters, TVs, and other technological triumphs.

First of all, here's a watery analogue of electricity, to dismay purists. A water pump sends water under pressure through a pipe to turn a waterwheel, which spins a saw. With most of its pressure

ELECTRICAL UNITS

spent, the water then flows back to the pump, where its pressure is boosted. Back it goes to the waterwheel, to do its work again, over and over.

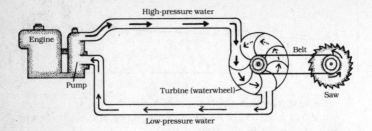

The pump is analogous to the electric generator (see GENERATORS AND MOTORS, ELECTRIC) in a power plant. The water is analogous to electricity in a wire. The waterwheel represents a user of electric current, such as an electric motor, a lamp, or a toaster. The pipes are water conductors: they represent electrical *conductors* — electric wires — and it's obvious that we need two of them, so that the stream of water (the stream of electrons is called an *electric current*) can be used over and over again (see ENERGY for more on electrical energy).

Now let us review some electrical units, named in honour of scientists important in electrical research.

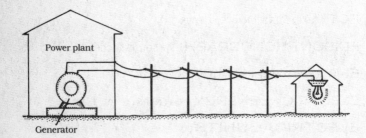

Volt (Alessandro Volta, Italian, 1745–1827) A unit of electrical force. Clearly, the stronger the force, the more power it can deliver to the user. (A stronger stream of water pushes harder against the blades of the waterwheel. But just as clearly, the stronger the force, the stronger the shock it can deliver. In Britain, electric current is delivered to houses at 240 volts. In the United States it's about 115 volts for user-accessible gadgets such as lamp sockets and about 230 volts for less-prone-to-fiddling devices such as electric cookers.

Ampere (André Marie Ampère, French, 1775–1836) A unit of quantity of electric current, analogous to the flow of water (e.g., litres per second). A small black-and-white TV takes about 0.5 ampere, or roughly 3 trillion (3×10^{18}) electrons, per second.

Watt (James Watt, Scottish, 1736–1819) A combination unit, to express what you're really buying, electric power. An electron receiving high voltage conveys more power than one with low voltage. Many electrons per second convey more power than few electrons per second. So we take account of both force and quantity — voltage and amperage — by multiplying the two. The resulting units are called watts. Thus a 240 volt kettle that draws 10 amperes is consuming 2400 watts. For the big stuff, such as electric cookers, we use a more convenient unit, the kilowatt, or 1000 watts.

Kilowatt-hour An electric oven doing its job on a Christmas turkey and assorted trimmings may be drawing a power of 9 to 10 kilowatts — but for how long? The more time, the more electricity used. The electricity board charges you for *power* in relation to the length of *time* you used it: kilowatts times hours, or kilowatt-hours (kwh). That figure, times the cost per kilowatt-hour, is what you pay.

Ohm (Georg Simon Ohm, German, 1787–1854) A unit of electrical *resistance*, analogous to the resistance in a water pipe. A narrow pipe offers more resistance to the flow of water than a wide pipe. This is also true of thin and thick wires. The filament wire in a 10-watt bulb has a high resistance because it's thin and is made of a high-resistance metal, tungsten. It allows few amperes to pass through and thus produces a feeble glow compared to the thicker tungsten filament of a 100-watt bulb, which allows ten times as many amperes to flow, producing a much brighter light.

Two more terms: *electric* and *electronic*. These are similar (both have to do with electrons) but not quite interchangeable. All devices that use electricity are electrical: light bulbs, doorbells, TV, neon signs, computers, toasters. However, some of these use electrons in special ways, not just as a stream flowing through a wire. For example, in a TV picture tube (CATHODE RAY TUBE) electrons form the picture. In a neon sign electrons cause a gas to glow. In a radio's SEMICONDUCTORS electrons operate tiny switches. These special devices and ways of using electrons are called electronic.

ELECTRIC BATTERY (CELL)   See BATTERY (CELL)

ELECTRIC CURRENT   See ENERGY

ELECTRIC GENERATORS AND MOTORS
See GENERATORS AND MOTORS, ELECTRIC

ELECTROCARDIOGRAPH

A device that measures and records electrical activity in the heart. The muscular tissue (*myocardium*) which encloses the four chambers of the heart, contracts and relaxes about 70 times per minute. This pumping action circulates blood throughout the body. Like any working muscle, the myocardium produces minute electrical currents, on the order of a few ten-thousandths of a volt.

A heartbeat begins with the contraction of the upper two chambers, the *atria*, of the heart. The contraction spreads, wavelike, to the lower two chambers, the *ventricles*. They contract, forcing the blood out of the heart into large blood vessels. The myocardium relaxes for less than half a second, and a new cycle of contraction begins. Each of these events produces a pattern of electrical currents. First, as the atria contract, a current flows downwards across them and disappears. Then, as the ventricles contract, a current flows from them towards the atria and disappears.

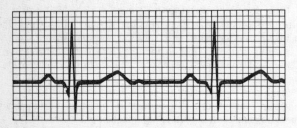

The currents pass through the body and reach the skin. Leads (wires) from the electrocardiograph end in electrodes, which are attached to the skin. The mechanism senses the currents and converts the flows, stops, and reversals of current to patterns of lines printed on graph paper, the *electrocardiogram* (ECG). To the trained eye, the ECG points up timing faults, contraction abnormalities, other heart problems, or, most often, no problems at all.

See also CIRCULATORY SYSTEM.

ELECTROCHEMISTRY See CHEMISTRY

ELECTROCONVULSIVE THERAPY (ECT)

A method for treating severe mental disease, especially depression, by inducing convulsive seizures in a patient. The most common method is to administer a brief (up to half a second) electric shock to the anaesthetized patient; in some cases, drugs are used to produce the convulsions. Usually several treatments are given over a period of time.

ECT is a controversial technique. It is often dramatically fast and effective, yet there is concern about its long-term effects on the brain's functioning and especially on memory. It is not known why these convulsions should produce beneficial results.

ELECTRODE *See* BATTERY (CELL)

ELECTRODIALYSIS *See* KIDNEY DIALYSIS

ELECTROENCEPHALOGRAPH

This spelling quiz special comes from Greek and Latin words for 'electrical writing of the brain', and it refers to the machine that records the brain's electrical activity — an activity that never stops, awake or asleep. The electroencephalograph senses the minute electrical currents (on the order of one hundred-thousandth of a volt) through electrodes placed around the patient's head. The currents are amplified and recorded as sets of waves traced on a moving strip of paper, the *electroencephalogram* (EEG). The expert can detect several kinds of rhythmic patterns in these *brain waves*. For example, the most prominent, called *alpha waves*, occur between 8 and 13 times per second. Abnormal patterns of waves assist the specialist in diagnosing epilepsy, tumours, and other brain disorders.

ALPHA WAVES

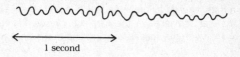

1 second

Other kinds of diagnostic equipment are also based on electrical activity in body organs. The best known is the ELECTROCARDIOGRAPH and the ELECTROMYOGRAPH (Greek *myo*, muscle), used in exploring certain kinds of muscle and nerve disorders.

ELECTROLYTE *See* BATTERY (CELL)

ELECTROMAGNETIC FIELD *See* FIELD THEORY

ELECTROMAGNETIC INDUCTION
See TRANSFORMER

ELECTROMAGNETIC RADIATION
See RADIANT ENERGY

ELECTROMAGNETIC SPECTRUM

Light, X-rays, radio waves, and several other kinds of energy are transmitted in the form of electromagnetic waves of various lengths. The entire range of these radiations is called the electromagnetic spectrum.

See also RADIANT ENERGY.

ELECTROMAGNETS

Devices that produce magnetism by means of a current of electricity. Not surprisingly, such devices are called *electromagnets*. You use them throughout your day in the motors that power your electric clock, food processor, office lift, air conditioner, and other appliances. In most motors the electromagnets receive an on-off, on-off (alternating) current 50 times per second (50-hertz AC). Each on-off cycle produces a small jolt of magnetism that causes a part called a *rotor* to rotate.

Electromagnets are called *temporary magnets*. The magnetism they produce is temporary because it depends on the flow of electricity. Turn off the current, and the magnetism is gone. This temporary quality is useful, for example, in the operation of your telephone. The earpiece of the phone contains an electromagnet that receives electric current from a distant telephone. The current flows in little bursts, caused by the vibrations of a distant speaker's vocal cords. The electromagnet in the earpiece causes a metal disc to vibrate in step with the speaker's voice, and that's what you hear as speech.

The most spectacular use of electromagnets is in the propelling mechanism of a PARTICLE ACCELERATOR. For a closer look at electromagnetism, *see* GRAND UNIFIED THEORY.

Another kind of magnet, the *permanent magnet*, requires no current. Your introduction to this kind may have come via a toy horse-shoe magnet or a magnetic compass. Permanent magnets in the doors of some cupboards and refrigerators keep the doors closed without the need for latches.

ELECTROMYOGRAPH   See ELECTROENCEPHALOGRAPH

ELEMENTARY PARTICLE PHYSICS

ELECTRON

The negatively charged part of an atom surrounding the positively charged nucleus. The number of electrons per atom determines the chemical characteristics and some other aspects of a particular element. Electrons can be forced to move from atom to atom through a conductor, such as a copper wire, producing an electric current.
 See also ELEMENTARY PARTICLES.

ELECTRON MICROSCOPE *See* MICROSCOPE

ELECTRON VOLT

A measure of electrical energy. It is the energy gained by one electron when propelled by a force of 1 volt — a tiny amount indeed. Many thousands of electron volts are contained in the spark that tickles the fingertip when one shuffles across a carpet and touches a doorknob. In a TV set, each electron that contributes to forming an image in the picture tube (and there are thousands of billions per second) carries a charge of 20,000 electron volts.

The term *electron volt* is used most often in describing the power of a PARTICLE ACCELERATOR (once called an atom smasher). The energyyy in these enormous machines is measured in larger units: MeV (million electron volts) and BeV (billion electron volts). You will also come across GeV, the G of which stands for *giga-*, pronounced 'jig-a' or 'gig-a' (from the Greek *gigas*, giant), which is replacing BeV. This is because a British billion means one million million (1,000,000,000,000) whereas an American billion is one thousand million (1,000,000,000). One million million (one billion) electron volts, is the output of the Fermilab atom smasher, the Tevatron, at Batavia, Illinois, which went into action in July 1983.

ELECTROPHORESIS *See* ANALYSIS

ELECTROSTATIC FIELD *See* FIELD THEORY

ELECTROWEAK FORCE

A composite force based on the electromagnetic and weak nuclear forces.
 See also GRAND UNIFIED THEORY.

ELEMENTARY PARTICLE PHYSICS *See* PHYSICS

ELEMENTARY PARTICLES

If you chopped a piece of copper into tiny particles, you would have ... tiny particles of copper. And if you chopped those, and the resulting pieces, again and again? Common sense and experience say copper, all the way down. Some kind of anti-common sense led the Greek philosopher Democritus (460?–370? asbc) to argue otherwise; he believed that copper and all other materials are made of tiny, unalterable, absolutely indivisible particles that he called *atoms* (Greek *a*, not + *tomos*, to cut). Democritus conjectured atoms of copper, of iron, of air, of water, and so on, but could offer no physical proof of his hypothesis. It faded from view until revived by the English scientist John Dalton (1766–1844), whose precise chemical experiments and measurements led him to the theory that there are indeed ultimate minimum particles — atoms. These later proved to be further divisible. The New Zealand physicist Ernest Rutherford (1871–1937) demonstrated that atoms consist of two basic structures:

1. A central part, the *nucleus*
2. A set of *electrons* (previously discovered) orbiting the nucleus

Electrons, to this day, have resisted further division; they seem to be truly elementary particles. The nucleus, however, has turned out to be a Pandora's box of subparticles. In 1911 Rutherford reported that the nucleus is composed of electrically charged particles, *protons* (Greek *protos*, first). In 1932 his former student, James Chadwick (1891–1974), identified some noncharged (neutral) particles in the nucleus and named them *neutrons*. Electrons, protons, neutrons — for a number of years this trinity seemed to describe the ultimate simple structure of the universe.

Ah, deceptive simplicity! The coming of the particle accelerator, which split the nuclei of various atoms, brought forth, as expected, some protons and neutrons, but there also materialized many kinds of unexpected particles, with a wide, wild variety of dimensions, masses, life expectancies, and behaviour styles. Over the years since 1932 more than 200 kinds, plus the original trinity, have emerged from the particle accelerators, clamouring for names and identities. They have been classified in many ways.

Classifying by mass This early system of grouping has turned out to be only partially relevant. Nevertheless, the following weighty terms have hung on, even though their original meanings have changed somewhat.

Leptons (Greek *leptos*, fine, small, light) A familiar example is the electron. Less familiar examples are the *muon*, three kinds of *neutrino*, and their antiparticles (*see* ANTIMATTER).

Mesons (Greek *mesos*, middle) There are no familiar examples, but here are three unfamiliar ones: *pion, kaon,* and *psi* particles, and their antiparticles.

Baryons (Greek *barus*, weight) Two familiar examples are the neutron and the proton. These two, because they are regular components of the nucleus, are called *nucleons*. Less familiar baryons are *lambda* and *sigma* particles (called *hyperons*) and their antiparticles.

Classifying by charge   Electrical charge shows up in electrons in the form of electric sparks, electric current, and lightning flashes. Other elementary particles also possess charge, either like an electron's, which is a negative charge (−), or like a proton's, a positive charge (+). Some particles (such as neutrons) show neutral (zero) charge.

Classifying by spin   The subatomic world of elementary particles is a fantastically busy place: everything is spinning (rotating) and/or circling (revolving) and/or vibrating, all at incredible rates, in millions and thousands of millions per second. These rates are quite specific (*see* QUANTUM THEORY).

Classifying by job description   There are four kinds of particles whose principal job is to transmit force from place to place (*see* GRAND UNIFIED THEORY). All of them are called *gauge bosons*. *(Boson* is in honour of Satyendranath Bose, Indian mathematician, 1894−1974).

Gravitons   Not yet discovered, believed to transmit the force of gravity between you and the earth. Without them you would sail off into space every time you took a step. (*See* GRAVITATION for the rest of the job description.)

Photons   Transmit the electromagnetic forces, including radio waves, visible light, and X-rays (*see* RADIANT ENERGY).

Intermediate vector bosons   Transmit the *weak nuclear force* inside atomic nuclei. These bosons are also called W+, W−, and $Z°$ particles, for less than earth-shaking reasons. The +, −, and ° refer to electrical charge.

Gluons   Transmit the *strong nuclear force* between the particles of which protons and neutrons are made.

And there — aha! — we have let the cat out of the bag. The question that was implied somewhere earlier was, are neutrons and protons indivisible, ultimate particles that cannot be divided into still lesser particles? And the answer is no, they are not fundamental. In 1964 Murray Gell-Mann described neutrons and protons as being made of *quarks*. (He had been intrigued, on reading James Joyce's *Finnegans Wake*, by the phrase, 'Three quarks for Muster Mark'.) There are six 'flavours' of quarks. (By now you may be sufficiently sensitized to physicists' whimsy to understand that *flavour* here has nothing to do with chocolate.) These flavours, to tease you further, have been labelled *up, down, strange, charm, top*, and *bottom*.

Quarks, thus far in history, are believed to be truly fundamental. Experiments and calculations have shown that a proton is assembled from two up quarks and a down, while a neutron consists of two downs and an up. The other particles are likewise built of quarks. Each flavour of quark, furthermore, come in three 'colours', red, green, and blue (having nothing to do with visual colour, of course). Six flavours of quark, then, each in three colours, come to 18 different quarks.

Is that it? Are these 18 the ultimate basic particles of matter? No, because there are also 18 different . . . *antiquarks*! All particles, it is believed, have antiparticles, which are, so to speak, opposite numbers in antimatter. For example, mesons consist of a quark and an antiquark.

Basically, then, our physical world seems to be constructed of quarks (six flavours, each in three colours, and their antiquarks), making up protons, neutrons, and other nuclear particles. These bundles of quarks, when surrounded by orbiting electrons, form atoms and molecules. The quarks are laced in place by gluons.

So the picture doesn't seem too complicated (we hope). Furthermore, as a reward for having ploughed through this far, we offer the following good news:

'Our physical world', mentioned earlier, refers not only to you, church steeples, and similar relatively durable objects but also to the momentary particles that appear and disappear in a billionth of a second, during the processes of radioactivity, nuclear fission and fusion, atom-smashing, and similar fleeting occurrences. If we exclude all the evanescent (but scientifically significant) particles, we are left with a much shorter list, to wit:

- *Two* kinds of quarks: up and down, in three colours, and their antiquarks
- *Two* kinds of leptons: neutrinos, electrons, and their antiparticles

What a relief!

ELEVATOR

In aerodynamics, a movable control surface at the tail of an aeroplane or glider, used to tilt the tail of the aircraft upwards or downwards.
See also AERODYNAMICS.

ELLIPTICAL GALAXY   See GALAXY

EMBOLUS   See STROKE

EMBRYO   See AMNIOCENTESIS

EMBRYOLOGY   See BIOLOGY

$E = mc^2$ See RELATIVITY

EMPENNAGE

The assembly of control surfaces at the tail of a glider or aeroplane.
See also AERODYNAMICS.

EMPHYSEMA AND BRONCHITIS

We begin with a short anatomical digression: air passes into and out of the lungs through a pair of wide tubes, or *bronchi* (Greek *bronchos*, throat). Each bronchus branches repeatedly (in fact, the whole system is sometimes called the bronchial tree), forming smaller and thinner tubes; the smallest are the *bronchioles*, and each of these ends in a cluster of tiny balloonlike air sacs, or *alveoli* (Latin *alveus*, hollow, cavity). The wall of an alveolus is only a single cell thick, and it is in contact with a net of equally thin blood capillaries. Together the alveoli and their associated capillaries provide a surface of more than 56 square metres (about 600 square

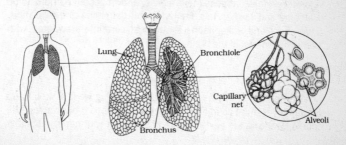

feet) in which oxygen from fresh air enters the blood and waste carbon dioxide leaves it.

Bronchitis is an inflammation of the bronchial system; it may be acute (of fast onset and short duration), as with a cold, or chronic (long-lasting). Chronic bronchitis, a serious disease, is becoming more prevalent and is often associated with emphysema.

In emphysema the normally elastic walls of the alveoli become flabby and stretched (Greek *emphysan*, to inflate). They break down progressively, reducing the area available for the exchange of gases. The victim feels short of breath and tires easily because the body cells are short of oxygen while suffering an excess of carbon dioxide.

Although the cause of bronchitis and emphysema is not known, outside chemical factors — air pollution and occupational exposure — are clearly involved in a large number of cases. Cigarette smoking, especially, is a major factor: emphysema is found 11 times as often in smokers as in nonsmokers.

ENDOCRINE GLAND *See* HORMONE

ENDORPHINS AND ENKEPHALINS

A group of natural morphine-like substances produced by the brain. Two forms of enkephalin have been found and several endorphins. They all have pain-relieving properties and are thought to be involved in the regulation of pain sensation.

They were discovered in the 1970s after it was realized that opiate drugs, like morphine and heroin, act on certain circuits in the brain. It was found that these same circuits are naturally affected by the body's own 'opiates'. The enkephalins were subsequently found throughout the nervous system where it is thought they act as *neurotransmitters* (substances that pass on nerve impulses between nerve cells). Endorphins, which seem to be concentrated in the pituitary gland, may have a hormonal role.

It has been suggested that acupuncture may work by causing the release of enkephalins or endorphins which suppress pain. Also, that vigorous exercise induces their release, allowing 'pain barriers' to be overcome and leading to a feeling of mild euphoria and relaxation. Endorphins and enkephalins are almost certainly associated with 'pleasure centres' in the brain.

ENERGY

Can be defined briefly as the capacity for doing work. A wound-up windup toy is more charged with energy than when it has run its little course, awaiting the next infusion of energy at your hand. A boulder

ENERGY

poised at the edge of a cliff has the capacity for doing a great deal of work, albeit of a destructive nature.

In each case, we are observing energy in transition from one state to another: from the potential state (Latin *potens*, having power) to the kinetic state (Greek *kinetos*, moving). Think of it as money in the bank turning into money being spent.

Mechanical energy A hammer in action is a huge mass of molecules — steel head and wooden handle — all engaged in a single concerted motion. Molecules in motion possess mechanical energy (Greek *mekhanikos*, machine, contrivance). When the hammer head strikes a nail, the mechanical energy is passed along to the nail, which moves. A rubber band, too, is a large mass of molecules. Stretch a rubber band and release it. The vibrating twanging band delivers a special form of mechanical energy, called *sound*, to the air. In fluids (liquids and gases) the molecules aren't bound together like the molecules of a steel hammer head or a rubber band, but they, too, have mechanical energy when they move together. A stream of water strikes the blades of a waterwheel. A stream of air strikes the blades of a windmill. The mechanical energy in the moving streams is passed along to the blades, and they move.

Heat energy We've seen that masses of molecules in concerted motion exhibit mechanical energy. Now let's look at the separate motions of individual molecules. Place your palm against your forehead. Palm and forehead probably feel equally warm, or nearly so. Then rub your hands briskly together for half a minute. Again, place a hand against your forehead and observe that this time the hand feels warmer. The molecules of your hand, your forehead, and of everything else in the universe are in motion all the time. They jiggle back and forth, up and down, colliding and rebounding constantly. This endless random motion is called heat. When you rubbed your hands together, you increased the rate of vibration. Faster vibration results in more heat.

Chemical energy Masses of molecules moving together exhibit mechanical energy. Separate molecules moving randomly create heat energy. Molecules are made of atoms, and the energy related to molecules and atoms joining, rearranging, and separating is *chemical energy*. An iron nail rusting away is a good example of chemical energy. The factory-fresh grey nail gradually takes on a reddish colour. A chemical change is taking place, in which atoms of iron in the nail combine with atoms of oxygen from the air to form molecules of iron oxide, or rust. This chemical change is accompanied by a discharge of heat energy.

ENERGY

Every chemical change is also an energy change – energy is either given off or taken in. From our viewpoint as living things, the most important chemical changes are oxidation and photosynthesis.

Oxidation In oxidation, oxygen is combined with another substance and heat energy is given off. The metabolism of foods by living things is an oxidation process. So is the burning of coal, oil, and other fuels and the rusting of metals.

Photosynthesis Photosynthesis (from the Greek for 'light' and 'put together') takes place in all green plants. These plants use the energy of sunlight to combine water and carbon dioxide from the air, forming sugar and similar energy foods such as starch. As a leftover, oxygen is released into the air.

So far we've dealt with whole molecules and atoms. The parts of the atom called electrons bring us to still another form of energy.

Electrical energy A copper wire consists of copper atoms. Each copper atom has 29 electrons whirling around a central portion, the *nucleus*, which contains 29 *protons*. A single proton can hold one electron in place. Thus 29 protons can hold on to 29 electrons; we say the atom is electrically balanced.

Suppose we force an extra electron, number 30, into a copper atom at the end of the wire. The extra electron disturbs the balance of the nearest atom, which reacts by forcing one of its own electrons into the next nearest atom, and so on, down the length of the wire. Actually, you pay the electricity board for doing exactly this – forcing electrons through wires. The electric generator sets up a stream of electrons from atom to atom. The stream is an *electric current*. When you switch on a 100-watt bulb, about 300 trillion (300×10^{18}) electrons flow through the wire in the bulb every second (*see* ELECTRICAL UNITS).

We have looked at energy forms, down the scale of size, from

- whole packages of molecules moving together (mechanical energy) to
- individual molecules in random motion (heat energy) to
- molecules and their parts – atoms – joining and separating (chemical energy) to
- parts of atoms – electrons – moving in a current (electrical energy).

Smaller and smaller particles, but believable. With instruments of increasingly higher power, we can keep track of these increasingly tiny phenomena.

To go further, we have to give up our belief in common sense. We enter a field populated with ghostly particles that have no dimension and no mass, that exist only when travelling at the speed of light, that disappear when they stop, and that can exist anywhere from billionths of a second to billions of years. These particles are the basis of radiant energy.

Radiant energy The ghostly particles are called *photons*. They result from events within the atom. Every atom has a nucleus and one or more electrons. Each electron moves around the nucleus, but it may occupy any one of several specific orbits, depending on its energy level. (Imagine Mercury, the planet closest to the sun, appearing suddenly in the orbit of Venus, or even farther out.)

An electron with the least energy (called *ground state*) occupies the orbit closest to the nucleus. An electron with greater energy is in *excited state* 1, then 2, and so on in farther-out orbits. When an electron loses energy, it drops from an orbit to a lower orbit and simultaneously emits one photon. The electron doesn't get smaller because photons have no dimension. The ghostly particles streak away at the speed of light because — and here the secret is out — they *are* light! A photon is a packet of light. It is commonly represented like this: ～～➤ the arrow-head indicates its direction of travel, and the waves indicate that a photon includes little trains of waves. A glowworm in action emits about 100 million photons per second; a small electric torch, about 100 thousand million.

Light is just a small section of the broad group called radiant energy (Latin *radiare*, to emit beams). All the members are alike in that they transmit energy by emitting wave-carrying photons. The difference is in their frequency (number of waves per second).

This drawing shows the frequency of some of the common forms of radiant energy. Notice the infrared frequency. These are the 'heat

ELECTROMAGNETIC SPECTRUM

Frequency (hertz, or cycles per second)		
10^{20}	Gamma rays	
10^{18}	X-rays	
10^{16}	Ultraviolet rays	Visible spectrum → Violet, Indigo, Blue, Green, Yellow, Orange, Red
10^{14}		
10^{13}	Infrared rays	
10^{10}	Radar waves	
10^{8}	TV waves	
10^{6}	Radio waves	

rays' whose photons make you feel warm in sunlight or near a hot stove. Notice also that the range, or *spectrum*, of visible light frequencies is only a very small part of the whole spectrum of radiant energy.

As noted, photons are emitted when an electron drops from an excited state to a less excited state. But what put it up to the excited state in the first place? Energy. A light bulb, for example, converts electrical energy to heat energy, which raises the energy level of electrons in the tungsten filament of the bulb. The electrons jump to higher orbits, immediately fall back, and emit photons, again and again, as long as the current continues, about 10 quadrillion (10^{24}) per second out of a 100-watt bulb. All start their ghostly existences at full speed, disappearing when they strike their targets: the ground, a plant, or the retina of your eye.

Another source of energy for creating photons is the most important of all: NUCLEAR ENERGY, the source of sunlight and starlight.

Photons have been described as 'accompanied by waves'. Waves in what? Not water, not air, for in fact, light waves travel most efficiently in a perfect vacuum – the emptiness of space.

ENERGY RESOURCES

According to a songwriter of the 1920s, 'The Best Things in Life Are Free' — referring to, among other things, love. To this, a physicist might add 'and energy'. A few minutes worth of the sunlight that shines on the earth contains enough energy to supply our heat, power, and transportation needs for an entire year. One teaspoon of tap water contains enough nuclear fuel to send a car across a continent. The energy is free — but the technology to extract it efficiently is yet to come. In the meantime, let's take a brief look at energy resources: nonrenewable, renewable, and unlimited.

Nonrenewable Coal, petroleum, and natural gas are *fossil fuels*, the remains of ancient plants and animals, distilled and concentrated by heat and pressure over millions of years. These nonrenewable resources fill the principal energy requirements for the world's industry and transportation, but they are also the principal renewable source of the world's political squabbles and military actions—such as the periodic upheavals of the great powers over access to the Middle East's oil reserves. Furthermore, the oil in these reserves may be depleted in perhaps half a century. We need to look to resourcs that are renewable and to others that are (comparatively) unlimited.

Renewable These are replaced as we use them and are to be had for the taking, except for the cost of constructing and maintaining the 'taking' machinery (a quite large exception). First place goes, of course, to sunlight, used directly or indirectly.

Solar energy *Photovoltaic systems convert* light directly into electric current. The light meter in a camera is a tiny, unassuming example. Spectacular examples are the panels (usually paddle-shaped) of solar cells affixed to some artificial satellites. They provide electric current for the electronic equipment on board. Both of these work on the photoelectric effect: The atoms of certain

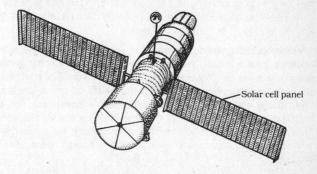
Solar cell panel

metals, such as selenium, possess electrons that are easily knocked out of place by light energy. In a photovoltaic system, a wafer of the electron-emitting metal is in contact with another metal that collects the electrons and passes them along into wires in a steady stream, while other electrons from the wires flow in to replaced them — a current of electrons, or electric current.

Solar heating This is another direct use of sunlight. In most solar-heating installations, the sunlight heats water flowing through blackened pipes (black is an efficient converter of light into heat). The heated water then flows into domestic hot-water systems, hot-water radiators, or swimming pools. And the consumer never receives a bill for the sunlight. Solar heating is becoming widespread, especially in places with plentiful year-round sunlight and comparatively advanced standards of living. In Israel, for example, about 60% of homes are equipped with solar heaters for hot water and heating.

Solar furnaces These use sunlight to heat water into steam that drives electric generators. One of these furnaces, in the Mojave Desert in California, is designed to provide enough power for 5000 homes. The sunlight is collected by more than 1800 heliostats (movable mirrors), each with an area roughly equal to that of the floor of a two-car garage. Computers control the heliostats to track the sun through the sky, reflecting its light onto a steam boiler.

Hydroelectric systems Electric generators driven by water turbines represent an indirect use of solar energy. The flowing, falling water that spins the turbine blades begins as rainfall from clouds. The clouds are formed from water vapour lifted by sunlight warming the earth. Totally renewable, and without the pollution produced by burning fossil fuels, hydroelectric energy is one of the ideal energy resources. Unfortunately, the number of places where such a system can be set up is limited. Very few sites in England and Wales are suitable, although there are over 50 plants in the more mountainous and wetter regions of northern Scotland.

Wind-electric systems Indirectly driven by solar energy, wind turbines (cousins of the picturesque windmill) drive electric generators. Winds are horizontal air currents caused by unequal heating of the earth's surface, and the source of the heat is, of course, the sun. Wind power shares the virtues of hydroelectric power: absence of pollution and total renewability, *if* strong, steady winds are available. That very large 'if' has kept wind-electric systems from becoming widespread and commercially important, although their use is growing.

Biomass There is a great deal of biomass (a general term for living matter and its organic by-products) in a jungle, less in a desert, almost none in Antarctica. It is a renewable source of energy — a familiar example is firewood; a not-so-familiar one, at least in the Western Hemisphere, is the dried dung of cows, camels, and yaks. Dung is clean-burning, nearly odourless, and constantly renewable (to the owners of cows, camels, and yaks). Collected rubbish is another potential source of energy.

Methane Marsh gas, or methane, represents a more technically advanced exploitation of biomass. In nature this gas is produced through the action of bacteria on dead plant matter in marshes. Commercially, methane is produced by a similar bacterial process in enclosed steel containers, using farm manure as the principal raw material. Methane is odourless and its combustion products (carbon dioxide and water) are also odourless and nonpolluting.

Geothermal heat The earth's internal heat is a source of energy, permanently renewable, at least for the next 100 million years (which can be reasonably regarded as permanent). Enormous as this source of heat is, it gives us only about 1/5000 as much energy as we receive from the sun. Some geothermal heat is a leftover from the time our planet was first formed, but much heat is being produced continually by the breakdown of certain earth elements, particularly uranium, thorium, radium, and polonium. These elements undergo radioactive decay. Small portions of their atoms break away, forming smaller fragments, and energy is given off, some of it in the form of heat. A good deal of this radioactive process takes place in rock layers deep below the earth's surface, heating the rocks to hundreds of degrees Celsius, much higher than the boiling point of water. In some places the rock layers are in contact with underground streams or lakes, so huge amounts of steam and hot water are produced. These can be piped up to the surface for hot water supplies and for use in steam-driven turboelectric generators.

Hot-dry-rock technology In this new method for extracting geothermal energy from areas without underground water, water is pumped down deep wells into places where the hot rocks have many fractures. The water heats up as it flows through the fractures and is pumped up to the surface through a second well.

Geothermal energy has been developed on a commercial scale in only a few places: New Zealand, Italy, Iceland, and California. The California geothermal power installation is located in an area called The Geysers, about 145 kilometres (90 miles) north of San Francisco. In the early 1980s it was generating power at a rate of about 660,000 kilowatts. (An electric hob and oven together use about 12 kilowatts.)

Tidal energy The energy of the tides is free of pollution, free of fuel cost, totally reliable — and totally unused commercially, except in one place, on the Rance River in France. As sure as clockwork (much surer, in fact), the waters of the earth heap up twice a day in every ocean basin, and just as surely, twice a day, the waters flow away and heap up elsewhere, in a continual travelling tidal swell around the earth, energized by the gravitational pull and the centrifugal force of the sun-moon-earth system. Put something in the way of this tidal flow — a water turbine, for example — and you can transfer some of the energy from the moving water to the blades of the turbine, to turn an electric generator. Free, clean, quiet — but only on the Rance. A tidal power station was begun on Passamaquoddy Bay between the American state of Maine and the Canadian province of New Brunswick in 1936, but it was cancelled when half completed by a political squabble in the US Congress.

Unlimited Energy unlimited, or nearly so, was the bright promise of *nuclear fission* and is the bright promise of *nuclear fusion (see* NUCLEAR ENERGY).

ENGINEERING ACOUSTICS See ACOUSTICS

ENKEPHALINS See ENDORPHINS AND ENKEPHALINS

ENTROPY

> This is the way the world ends
> This is the way the world ends
> This is the way the world ends
> Not with a bang but a whimper

T.S. Eliot's lines report the end of the cosmological story. The beginning, according to most astrophysicists, was the big bang (see COSMOLOGY). Its ultimate end — who knows? — may be the last whimper on the way to maximum irreversible entropy, which might be defined as 'no further change'.

All the happenings of the universe, from the majestic rotation of galaxies to the infinitesimal vibration of electrons, are manifestations of energy diffusing into unavailability, falling towards maximum entropy. A newly formed raindrop, charged with energy by virtue of being high up in a cloud, falls and strikes a mountainside. The drop flings up tiny particles of soil, losing some of its energy in doing so. It continues downwards, flowing into a stream, and perhaps gives up a bit more of its energy to a waterwheel, turning an electric generator. Eventually it flows down to the sea, where it can fall no farther. Now, together with countless numbers of its fellow raindrops, it has reached maximum entropy, so far as energy available from falling goes.

That's not the end, of course. Another source of energy is available to lift the raindrop again, to recharge it with energy. That source is the heat from our local star, the sun. But the sun, and the billions of other stars that fill the universe, have a limited amount of energy to radiate. Eventually, one after another, they will cool down to dark masses, their energy scattered throughout space, no longer available. They will have reached maximum entropy.

And is this, then, the way the world ends? Perhaps not. Read about the oscillating universe under COSMOLOGY.

ENVIRONMENTAL ACOUSTICS See ACOUSTICS

ENZYME

An organic catalyst, that is, a catalyst made by a living thing. A catalyst is a substance that promotes a particular chemical reaction without itself being used up in the process.

The supreme achievement of catalysts is life. A plant or an animal is alive only as long as it continues carrying on a variety of chemical reactions. This must happen at moderate temperatures and at a rate thousands or even millions of times faster than the reacting molecules, left to themselves, would provide.

The feat is accomplished by enzymes — catalysts made within the cells of all living things. Early studies of these substances involved reactions in yeast cells, leading to the term *enzyme*, from the Greek words meaning 'in yeast'.

Thousands of chemical reactions take place in complex organisms — ourselves, for example: large molecules of food are broken down (digested), small molecules are built up to form body tissues, muscles contract and relax, sense organs respond to stimuli, and wastes are produced, collected, and excreted. These actions, and thousands more, are the product of enzyme-mediated chemical reactions. There are thousands of enzymes, each specific for its individual reaction.

The lack of an enzyme — call it X — slows or stops reaction X. Depending on the role of reaction X in the body's chemistry, that lack can produce disorders ranging from minor to fatal (*see* GENETIC DISEASES).

People long ago learned to profit from enzymes. The ancients fermented wine and leavened bread, unaware of the help they got from enzymes made by yeast cells. Among many uses, we employ enzymes to tenderize meat, remove stains (in washing powders), and make antibiotics. Enzymes also make possible laboratory tests to diagnose cancer, heart ailments, and other diseases; these tests are based on the fact that the body's output of certain enzymes changes in the presence of certain diseases.

Scientists are studying an enzyme produced by the white rot fungus, a common organism that lives on dead wood. The enzyme (it may be a group of enzymes) is *lignase*. It breaks up the molecules of *lignin*, an extremely tough constituent of wood. Lignase has now been found to attack and break up the molecules of many toxic and carcinogenic substances, such as DDT, so it may have potential for cleansing contaminated soil and water. It may also provide a method for making liquid fuels from low-grade coal (*lignite*) and for producing useful chemicals from agricultural waste.

EPSTEIN-BARR VIRUS *See* HERPESVIRUS DISEASES

EROSION OF ROCK *See* ROCK CLASSIFICATION

ESCAPE VELOCITY *See* SPACE TRAVEL

ESCHERICHIA COLI *See* GENETIC ENGINEERING

ETHANOL *See* ANALYSIS

ETHYL ALCOHOL *See* ANALYSIS

EUROPEAN LABORATORY FOR PARTICLE PHYSICS *See* PARTICLE ACCELERATOR

EUTROPHICATION

A process in which the supply of plant nutrients in a lake or pond is increased. The word is from Greek words meaning 'well-nourished', although 'overnourished' would be more accurate. In time (centuries or millennia), the result of natural eutrophication may be dry land — from plant overgrowth — where water once flowed.

Springs and rivers that drain into lakes carry dissolved nitrates, phosphates, and other compounds — natural fertilizers — washed from the soil. The fertilizers stimulate the growth of algae and other water plants, and these provide food and oxygen for fish and other water-dwelling creatures. A lively establishment ensues and grows. As more nutrients arrive, the exuberance of plant growth may produce overcrowding; plants die off, and a surplus of dead and decaying vegetation depletes the lake's supply of dissolved oxygen; fish begin to die off. The accumulating dead plant and animal material changes the deep lake to a shallow one, then to a swamp, which in turn gives way to dry land.

Eutrophication has been speeded up enormously by human activities. Fertilizers from farms leach into springs and rivers or are washed in on eroded soil. Sewage and some industrial wastes, even after treatment to remove solid materials, are a rich source of nitrates and phosphates, as are some detergents. Lake Erie in North America, the Sea of Galilee in the Middle East, and Lake Baikal in Siberia are far-flung examples of major lakes suffering from some degree of artificially accelerated eutrophication.

EVAPORATION *See* MATTER

EVAPORATOR *See* REFRIGERATION

EVOLUTION

From Latin *evolutio*, 'unrolling'. More than a million and a half kinds of plants and animals live on the earth today. There are more than 25,000 species of beetle in the United States alone. How did this enormous diversity of life arise? The scientific answers to that question are based on the work of Charles Darwin (1809–1882) and Alfred Wallace (1823–1913).

These two English naturalists, working independently, came to the same conclusion: the characteristics of living things are not fixed; they can and do change. Thus a group of closely related, interbreeding organisms (a species) can give rise to other species over a period of time. The idea that living things change (organic evolution) is not new; the Greek philosopher Aristotle (384–322 BC) was an early proponent. The great contribution made by Darwin and Wallace was in explaining how such change could occur.

In 1858 the two men collaborated in summarizing their independent conclusions. The next year Darwin published *On the Origin of Species*, in which he proposed the theory of *natural selection*. Its main points are:

- *Overproduction leads to a struggle for existence*. Plants and animals produce far more offspring than can survive on limited amounts of food in a limited space.
- *Offspring vary*. For example, in a school of young fish, a few may have slightly stronger tail muscles than the rest. The favoured few can swim faster.
- *The fittest survive*. The fastest swimmers are best adapted (most fit) to elude predators; on the average, they are most likely to survive longest and produce the most offspring. But the race for survival does not necessarily go to the swift. In a place where small openings in the rocks offer hiding places from predators, smallness may have greater survival value than speed. Nature (the environment) weeds out the least fit — a process Darwin dubbed 'natural selection'.
- *Variations are inherited*. Therefore, the traits that make for fitness continue. This was the weak point of the theory, because nobody knew how traits pass from one generation to the next. Today the mechanics of heredity are well understood, and they bear out the correctness of this part of the theory.

An example of evolution in action today involves some pathogenic (disease-producing) bacteria. They now pose a serious medical problem, having changed in a way that makes them resistant to antibiotics. A similar evolution in some insects has made them resistant to insecticides.

A large body of scientific evidence supports the concept of evolution:

- Fossils show the stages through which many plants and animals

have passed; the evolution of the horse from a four-toed, cat-sized creature is a good example.
- Thousands of studies of the structure, embryonic development, chemistry, and geographic distribution of organisms point to the descent of widely differing species from common ancestors.
- Chemical dating, radioactive dating, and other methods enable scientists to estimate the time that life has existed on earth (2000 to 3000 million years) and the times at which various organisms evolved — for example, the earliest fishes 500 million years ago, mammals 200 million years ago, the earliest apes 25 million years ago, and modern humans (*Homo sapiens*) 50,000 years ago.

In 1859 Darwin's work stirred up widespread opposition among people who felt that the concept of evolution violated their interpretation of the Bible. Similar views today have resulted in several American states passing laws that make compulsory the teaching of 'creation science' (also called creationism) in schools where evolution is taught. (In 1987 the US Supreme Court overturned the creation science law of Louisiana.) Some principles of creation science (from the Arkansas law) follow:

- The entire universe — its stars, planets, galaxies, plants, animals, and energy — was created all at once, out of nothing.
- Life on earth was created at some time between 6000 and 12,000 years ago.
- Since the creation, there may have been some *minor* changes in some of the originally created plants and animals.

An argument presented not in the state law but by creationists in reply to questions about fossils (e.g., dinosaur skeletons) goes something like this: these fossils do not indicate that such animals ever actually existed; the fossils may have been placed in the earth by the creator of the universe in order to test the faith of the true believers.

EXCITED STATE (OF ELECTRONS) See ENERGY

EXPANDING UNIVERSE

Astronomical observations indicate that the galaxies of the universe are receding from one another — that is, that the universe is expanding.
See also COSMOLOGY.

EXPERIMENT See SCIENTIFIC TERMS

EXPERT SYSTEM

A computer program or set of programs that provides 'expert' advice. An expert system, or *knowledge-based system* as it is sometimes called, contains a *database* of information relating to a particular subject, such as finance, medicine, or air-traffic control. The information, or 'knowledge', which is provided by teams of experts, is stored as a set of rules on which answers to problems are based.

For example, a doctor might key in a group of symptoms and receive a diagnosis and suggested treatment. Or an oil company could use an expert system to investigate areas for drilling.

These systems are designed to 'sense' a particular line of enquiry and ask appropriate questions. This can make them appear 'intelligent'. However this is not so. Their responses are based entirely on the information and rules programmed into them. The recommendations they give can therefore be wrong or at least less-than-perfect solutions.

An ethical question: who is to blame if the wrong action is taken on the basis of poor advice given by an expert system? The system designer? The supplier of the 'knowledge'? The person who trusted a computer to be infallible?

EXPLORER I SATELLITE   See VAN ALLEN BELTS

EXTRUSION   See MANUFACTURING PROCESSES

EXTRUSIVE FORMATION (GEOLOGY)
See ROCK CLASSIFICATION

FACTOR VIII See HAEMOPHILIA

FALLING STAR See SOLAR SYSTEM

FATS See CHOLESTEROL

FAULT

A break in a rock formation, caused by a shift in the earth's crust. A horizontal shift causes a horizontal displacement in the surface features, such as the sideways dislocation of a road or streambed. A vertical shift, if it lowers the downstream part of a riverbed, produces a waterfall; if it lowers the upstream part, the river piles up and rises against the newly formed wall, resulting in a pond or lake. A fault may be as small as a metre or two or larger than the 1000-kilometre (600-mile) horizontal San Andreas Fault in California. The 106-

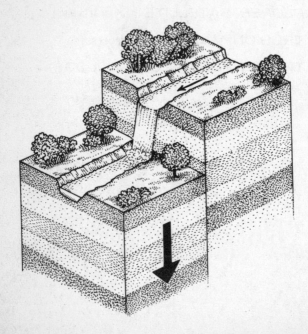

metre (350-foot) drop of the Victoria Falls in Africa is the result of a vertical fault. Rapid displacements are caused by earthquakes.

See also PLATE TECTONICS.

FERTILE FUEL   See NUCLEAR REACTOR

FERTILIZATION   See DNA AND RNA

FETUS   See AMNIOCENTESIS

FIBRE OPTICS

A system for transmitting light through hair-thin flexible rods (fibres) made of transparent glass or plastic. One familiar use of the fibres is in ornamental displays ('light trees'), but more important is their use in examining interior organs of the body and as economical substitutes for telephone cables.

Ordinary glass fibres transmit light quite well, with little loss along the way if the fibres are straight. Bend them slightly, and some of the light scatters sideways out of the fibres. A little more bend and so much light is scattered that they're useless as light transmitters. The special fibres used for medical and telephone-cable purposes have a coating, or *cladding*, of a different formula of glass that keeps the light from escaping. Shine a light through one of these fibres a whole kilometre long, with hundreds of bends and wiggles in it, and the light emerges at the far end with half its brightness still left.

Medical instruments using fibre optics are named according to the part of the body they are designed to examine: bronchoscope (bronchial tubes), cystoscope (bladder), gastroscope (stomach), and sigmoidoscope (lower large intestine). In all of these, a bundle of fibres transmits light from an outside lamp to illuminate the part of the body being examined, like a flexible torch. Another bundle of several thousand fibres has an objective lens at one end and an eyepiece lens at the other. The objective lens forms an image of the body part. This image is transmitted, point by point, through the fibres, to the eyepiece lens, which magnifies the image. The instrument is in effect a kind of interior-viewing microscope.

For telephone cables, glass fibres are used in place of copper wires. A pair of hair-thin glass fibres can carry several thousand conversations at one time, replacing several hundred wires within a cable as thick as your fist – an enormous saving in space and money. The telephone messages are first converted by a laser apparatus from electrical currents to pulses of light, are transmitted through the glass fibre, and are then converted back to electrical form at the far end. There they are sorted out and sent on to their receiving destinations. In Britain, all new telephone lines use fibre optics. Glass fibres are also used for high-density phone lines between major cities and for cross-Channel cables (e.g. to Holland). Eventually, most copper cables will be replaced by glass fibres.

FIBRIN See HAEMOPHILIA

FIELD THEORY

Physicists' attempt to describe (not altogether to their own satisfaction) a force acting at a distance — that is, through empty space — without an intervening substance to transmit the force.

Consider, for example, the space between a diving board and the water below (ignore the air, which isn't involved). If you walk onto the far end of the board, the space is altered — strained, so to speak — by the interaction between the earth's gravitational force, pulling down on you, and your own gravitational force, pulling upwards on the earth. This 'strained' condition in space is regarded as the *gravitational field* of the earth-you system. Similar fields are assumed in an earth-football system, the earth-moon system, the sun-earth system, the sun-Mars system, and so on.

The field theory applies to other forces besides gravity. If you were to rub a piece of plastic with a cloth, you would produce an *electrostatic field*. Electrons from the cloth are rubbed onto the plastic. Every electron possesses its own tiny electrostatic field. Crowded together, stationary (*static*), their fields add up to a larger electrostatic field. You can demonstrate its existence by holding the rubbed plastic near a bit of paper. The electrostatic field between the two objects, caused by stationary electrons, acts to attract the paper and the plastic towards each other.

Electrons *in motion* produce a *magnetic field*. In a permanent magnet, such as a horseshoe magnet, the field is generated by electrons *spinning inside* the atoms of the magnet. In an electromagnet (a temporary magnet), the field is generated by electrons *flowing from atom to atom* through wires (an electric current). Such a field is called an *electromagnetic field*.

Two more fields, neither of which you have ever experienced, are extremely important. These are the fields between the particles *inside the nucleus* of the atom and are involved with the production of nuclear energy.

See also GRAND UNIFIED THEORY.

FILTRATION See ANALYSIS

FIN

A fixed vertical surface at the tail of an aeroplane or glider, used to maintain the right-left stability of the aircraft.
See also AERODYNAMICS.

FIREBALL   *See* SOLAR SYSTEM

FISSION   *See* NUCLEAR ENERGY

FLAPS

On an aeroplane, movable control surfaces on the wing that are used to add lift or provide braking force.
See also AERODYNAMICS.

FLETCHER-MUNSON CONTOUR
See LOUDNESS CONTROL

FLIP-FLOP CIRCUIT   *See* COMPUTER

FLOPPY DISK

The most commonly used medium for storage of computer data. Floppy disks, which resemble gramophone records, are also known as *diskettes, floppies*, and *disks*. They are made of a thin, flexible plastic and are coated on both sides with a magnetic substance like that in a tape recorder cassette. Floppies are usually 3½, 5¼, or 8 inches (89, 133, or 203 millimetres) in diameter.

The disk, contained in a sealed envelope that protects its delicate surface, fits into a machine called a DISK DRIVE. The drive records ('writes') onto the disk and plays back ('reads') from it.

The floppiness is not a virtue; it is simply the result of the thinness of the plastic. A more rigid type of disk, called a *hard disk*, operates in its own kind of disk drive. It is much more expensive and not as portable as a floppy disk, but it does provide far greater storage capacity and more rapid access to the data recorded on it.

FLUID

From Latin *fluere*, 'to flow'. A substance that flows and takes on the shape of its container. Liquids and gases are fluids.

FLUTTER

A regular variation in the pitch of sounds made by a record or tape player. A long-playing (LP) record normally rotates 33⅓ times per minute. If the rate of rotation varies 6 or more times per second, the wobbly-sounding effect is called *flutter*. (Note, authorities differ on this — some say 6 variations per second, others 10, still others 20.) Below 6 (or 10 or 20) times per second the effect is *wow*. Some acoustic hairsplitters use two more terms: from 30 to 200 variations per second produce *gargle*, and above 200 is *whiskers*. Variations from normal speed in a tape player produce the same effects. Flutter and its companions are more noticeable in music than in speech and are especially bothersome when long-held notes are played.

FM *See* RADIO

FORTRAN *See* COMPUTER LANGUAGES

FOSSIL FUELS *See* ENERGY RESOURCES

FOURTH DIMENSION *See* RELATIVITY

FRAME OF REFERENCE *See* RELATIVITY

FREON *See* REFRIGERATION

FREQUENCY ANALYSIS *See* ANALYSIS

FREQUENCY MODULATION *See* RADIO

FREQUENCY OF ELECTRIC CURRENT
See RADIO

FREQUENCY OF SOUND WAVES
See DOPPLER EFFECT

FRONT *See* AIR MASS ANALYSIS

FUNDAMENTAL FORCES

The four forces — gravitation, electromagnetism, and the strong and weak nuclear forces — believed to be the basis of the universe's matter and energy.
See also GRAND UNIFIED THEORY.

FUNDAMENTAL TONE
See SOUND RECORDING; TONE CONTROL

FUSE

An electrical protective device based on the adage that a chain is only as strong as its weakest link. A fuse (Latin *fusus*, to pour or melt) is a weak link — a strip of metal that has a low melting point — inserted in the chain called an electric circuit. If, for some reason, too strong a current were to flow through the circuit, the wires might heat up and become a fire hazard. However, if a fuse were inserted in this circuit, the fuse would melt, leaving a gap and thus breaking the circuit. After the overload situation has been attended to, the melted fuse must be replaced to restore the current.

An octopus of plugs in a socket is sometimes used as a horrible example of overloading a circuit. Nothing of the sort. It may be unaesthetic, but overloading doesn't depend on the number of plugs or pieces of apparatus plugged in. It depends on the number of *amperes* (amps) flowing through the fuse or circuit breaker. Thus if it's a 13-ampere fused socket (fairly standard in home wiring) and you plug in both a 10-ampere kettle and a 12-ampere washing machine, you'll blow it. If you're using a desk lamp (about 0.25 amp), a radio (0.5 amp), a floor lamp (0.5 amp), and a fire (4 amps), however, you'll have a messy-looking socket, and a potentially dangerous assortment of leads, but everything will work.

See also ELECTRICAL UNITS.

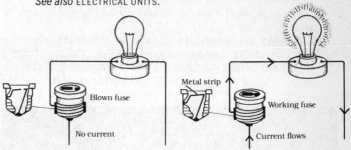

FUSELAGE

The central body of an aeroplane, to which the wings, tail, and (sometimes) engines are attached. It holds the passengers, crew, and cargo.

See also aerodynamics.

FUSION   See NUCLEAR ENERGY

GALAXY

A collection of stars, gas, dust, and (perhaps) planets, and other astronomical objects. On a dark, clear night a pale glowing band can be seen arching through the sky. Its resemblance to streams of milk earned it the name Milky Way, or galaxy (Greek *galakt*, milk). A pair of binoculars clearly shows that the 'milk' is really thousands of dim stars. In fact, the Milky Way, the galaxy in which we live, comprises millions of stars, among them the sun, drawn together by gravitation into a vast, vaguely disc-shaped conglomeration. It is classified as a *spiral galaxy*, because arms trail outwards in a spiral from and around the disc's centre. In one of the arms lies our sun, with its solar system in gravitational tow. From our viewpoint at X in the galaxy, we see many more stars — the 'milk' — when looking along the arm towards A than we see by looking in the direction of B. Most galaxies have no arms and are classified as *elliptical galaxies* or *irregular galaxies*. Everything the naked eye sees in the sky (with a couple of exceptions we'll note later) is a part of the Milky Way.

Galactic distances and sizes are measured in LIGHT-YEARS. One light-year is approximately 9.5 billion kilometres (6 billion miles). The Milky Way is about 100,000 light-years in diameter; the sun, positioned in one of the arms, is about 30,000 light-years from the centre of the galaxy.

The Milky Way has neighbours. One, the Andromeda galaxy, is a spiral galaxy much like ours, but with nearly twice the diameter. It is about 2 million light-years away. It, and the Magellanic Clouds, two smaller irregular galaxies only 200,000 light-years distant, are the

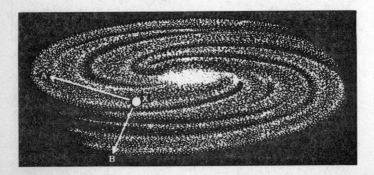

only objects outside our galaxy that you can see with the naked eye. Now to expand our horizons:

- The Milky Way, the Andromeda galaxy, and the Magellanic Clouds, along with some two dozen other 'nearby' galaxies, are members of a cluster called the Local Group. Local? Yes, if you consider that the nearest other such cluster to us is about 60 million light-years away.
- Markarian 348, the largest known galaxy, is 1.3 million light-years in diameter and some 300 million light-years from us.
- Some clusters hold thousands of galaxies.
- Thousands of clusters are known to exist, all moving away from one another at enormous speeds.

See also COSMOLOGY.

GAMMA RAYS

RADIANT ENERGY of extremely short wavelength and great penetrating power. Substances undergoing radioactive decay emit gamma rays, among other forms of energy.

GARGLE See FLUTTER

GAS See MATTER

GAS-COOLED REACTOR See NUCLEAR REACTOR

GASOHOL See SYNTHETIC FUELS

GATE, ELECTRONIC See BOOLEAN ALGEBRA

GAUGE BOSON See ELEMENTARY PARTICLES

GENE See DNA AND RNA

GENE MAPPING See GENETIC DISEASES

GENERATORS AND MOTORS, ELECTRIC

The song title 'Is That All There Is?' could apply to the workings of electric generators and motors. These basically simple machines have been around for a century and a half. Of course they have been vastly altered and improved, but they are essentially the same machines as their humble ancestors. In fact, they are essentially the same machine: Any electric generator can work as an electric motor, and vice versa.

With all that enticement, you may wish to go further: There are only two basic parts — one that rotates, called a *rotor*, and one that doesn't rotate, that remains stationary (static), called a *stator*. (There are other names for these parts — armature, field coil, and so on — but let's stay with the names that describe the functions, rotor and stator.)

Here's a crude but basically honest diagram of the two parts working as a generator. The stator is a coil of wire (connected to a light bulb, to use the generated electricity). The rotor is a bar magnet (most machines use an electromagnet, but the bar magnet is easier to understand). The rotor is pivoted so that it can be spun (rotated) by hand, windmill, water power, steam or petrol engine, or what have you.

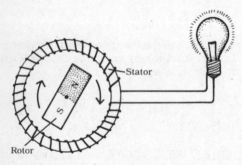

(The next paragraph can be skipped without doing mayhem to the train of thought, but it's an interesting sidetrack.)

Notice the *N* and *S* marks on the magnet. If you balanced the magnet on a thread, it would be a compass: One end (pole) would point to the earth's magnetic north pole. This north-pointing pole is labelled *N*. The same is true of the *S* pole — it points to the earth's magnetic south pole. The important fact is that *the two poles of the magnet work in opposite ways*. The *N* pole will attract another magnet's *S* pole; likewise, the *S* pole will attract another magnet's *N* pole.

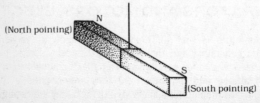

GENERATORS AND MOTORS, ELECTRIC

In the same way, the opposite poles of a *moving* magnet have opposite effects on the electrons in a coil of wire. One pole will drive electrons forwards; the other pole will pull electrons backward. So to run the generator, we spin the rotor. At each full turn (cycle) of the rotor, first one pole sweeps across the stator coil, driving electrons forwards; then the other pole sweeps across, pulling electrons backwards. Even though the rotor is spun in one direction, the electrons flow forwards and backwards, cycle after cycle, *alternating*. If you could spin the rotor 50 times a second, you would be generating a 50-hertz (50 cycles per second) *alternating current*, or *AC*. This is the kind supplied by the electricity boards in Britain and in many parts of Europe.

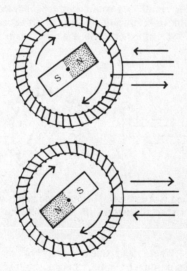

(An interesting variation on this theme is the use of a hot liquid or a gas in place of a coil of wire; see MAGNETOHYDRODYNAMICS.)

This is a diagram of three cycles of an alternating current. Above the line electrons flow in one direction; below the line, in the opposite direction.

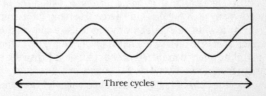

A brief summary: in a generator, we *cause* a magnetic rotor to spin. We supply an *input* of mechanical energy (water power, etc.) and receive in return an *output* of electrical energy — alternating current (AC).

Now let's convert the generator into a motor. That is, let's supply an *input* of electrical energy and receive in return an *output* of mechanical energy. Nothing to it; just connect the coil of wire (the stator) to a source of AC. As the alternating current flows through, it sets up an alternating magnetic field around the coil. First one end of the coil becomes an *N* pole, and the other an *S* pole. Then the polarity is reversed, and then reversed again. So the magnetism of the coil, the stator, keeps reversing, but the rotor's magnetism is unchanged. The result is a continuous repel-attract-repel action, causing the rotor to keep spinning, first by attraction and then by repulsion, half-cycle after half-cycle, again and again.

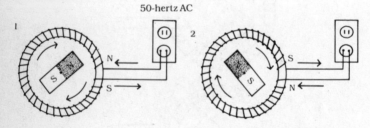

That's all there is. Yes, truly. The rest is variations on the theme — adaptations for special purposes. If you're still with us, here are a few:

Variation 1. What about direct current? A car's starter, for example, works on direct current, supplied by the battery. And a car's starter is basically an electric motor. An ingenious device called a *commutator* (Latin *commutare*, to change) keeps reversing the direction of the current to produce the repel-attract-repel action we saw with AC, to keep the rotor turning.

Variation 2. Can a generator produce DC? The basic, primeval generator, as you saw, is an AC machine, because its rotor's *N* and *S* poles drive electrons first in one direction and then the other. Yet there are many uses for DC — to charge a battery, for example. One solution is, as with the DC electric motor, to use a commutator on the generator. It acts to reverse the direction of half the output, resulting in a pulsating direct current. A second solution is to use the primeval electric generator and send its AC output into a *rectifier*, a device that allows the current to flow in one direction only. The combination

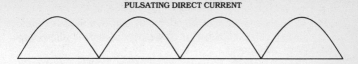

PULSATING DIRECT CURRENT

puts out a pulsating direct current; the ALTERNATOR in a car is used in such a combination.

Variation 3. There are many uses for electric motors with varying speeds — for example, electric fans, blenders, food processors, electric trains. Such speed control is achieved by regulating the strength of the electric current, usually with a device called a *rheostat* (Greek *rheo*, current), or variable resistor. In this diagram, the sawtoothed lines represent resistance wire. Such wire is made of a metal that is a somewhat poorer conductor of electricity than copper. The longer the wire, the greater the resistance; therefore, the weaker the current that reaches the motor and the slower it goes.

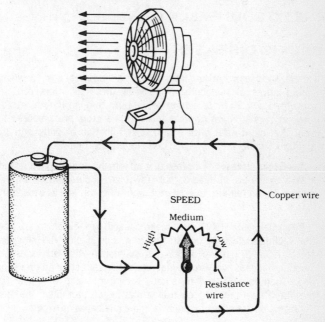

GENERIC DRUGS

What's the difference between the tranquillizer Valium and the tranquillizer diazepam? Answer: the uppercase *V* and the lower case *d*. Valium is the registered brand name of the drug whose *generic name* is diazepam. Brand names almost always cost more than generic drugs, often many times as much. In their effects there is usually no difference.

The British National Formulary lists drugs and medicinal preparations that may be prescribed and dispensed in Britain. From 1986, government restrictions on certain categories of preparations, such as tonics and mild tranquillizers, has limited the range of drugs that can be prescribed under the National Health Service. In particular, this applies to brand-name drugs where a generic one is available, the aim being to reduce the national drugs bill by eliminating duplication. So far, an estimated £75 million per year has been saved.

GENE SPLICING   *See* GENETIC ENGINEERING

GENETIC COUNSELLING   *See* GENETIC DISEASES

GENETIC DISEASES

Hereditary diseases arising from defects in the genes, also known as inborn errors of metabolism. Many of these disorders can be diagnosed by *genetic screening* before birth through the process of AMNIOCENTESIS. Knowing what to expect, doctors are prepared to treat the infant after birth or advise termination of pregnancy by abortion. Among the many such diseases are the following:

Tay-Sachs disease   A disease that affects the nervous system and is fatal by the age of three or four. Tay-Sachs occurs almost entirely among Jewish families of Eastern European origin and is caused by lack of an ENZYME.

Phenylketonuria (PKU)   A disease caused by lack of an enzyme that facilitates the elimination from the body of phenylalanine, an *amino acid*. The body needs this substance, but an excessive amount of it causes severe mental retardation. A test for an excess of phenylalanine can be done shortly after birth, and PKU can be prevented by a precisely controlled diet begun promptly. The diet meets the body's needs for phenylalanine but leaves no excess.

Sickle-cell anaemia   A disease that occurs mainly among black people. Normally, oxygen from the lungs is carried to all parts of the body by the protein haemoglobin in the disc-shaped red blood cells. In sickle-cell anaemia, some of the haemoglobin molecules are defective, causing the red blood cells to take on a hooked (sickle) shape; some of the sickle cells break up, and others clog small blood vessels, interfering with the circulation. Body tissues, deprived of part of their oxygen needs, are damaged. Most victims die before the age of 20.

Thalassaemia (Cooley's anaemia)   A disease somewhat similar to sickle-cell anaemia but found among people of Mediterranean ancestry. In this case, the oxygen supply is deficient because the defective haemoglobin causes some of the red blood cells to break up.

Cystic fibrosis   A disease of people of northern European ancestry in which the salivary glands, mucus glands, and other glands malfunction. In time, thick mucus accumulating in the air passages of the lungs interferes with breathing, leading to lung infections.

The increased understanding of genetic disorders and the development of tests for them have produced a new field of expertise — *genetic counselling*. The counsellor draws on a family's health and medical history and on diagnostic tests of the prospective parents and the fetus. From the data, the counsellor can calculate the probability that a particular disorder will appear in a child.

Recently genetic scientists have given a lot of attention to *gene mapping*, the work of nailing down the location and sequence of the genes. Today, only tiny fragments of the gene map are clear, like a map of Europe showing only Luxembourg and a 10-kilometre stretch of the Rhine.

Plans are being made to map the entire human gene system, or *genome* (somewhere between 50,000 and 100,000 separate genes), an enormously complicated and expensive undertaking. However, the complete map could be the first step in preventing and treating a wide variety of diseases.

See also DNA AND RNA; PROTEINS AND LIFE.

GENETIC ENGINEERING

A technique, also called *gene splicing* and *recombinant DNA technology*, whereby a section of DNA, comprising one or more genes, is removed from a cell and recombined with the DNA of another cell. The receiving organism is said to be *transgenic*.

Genes are the hereditary material that determines what a living thing will be — how it will develop and function — and what its descendants will be. Through recombinant DNA technology, scientists alter both the organism they work with and the generations that follow it. For example, genetically engineered bacteria now produce human-sequence insulin for use by diabetics.

GENETIC ENGINEERING

Juggling genes Much recombinant DNA work is done with bacteria, so a bit of background on them is in order.

In most organisms the DNA is contained in bodies called *chromosomes*. In bacteria, a small amount of DNA is also held in rings called *plasmids*, and it is these that are used in gene splicing. One often-used bacterium is *Escherichia coli (E. coli)*, an organism normally present in vast amounts in the human intestine.

Bacteria produce substances called *restriction enzymes*. Each enzyme is specific — a kind of molecular axe that cleaves a DNA

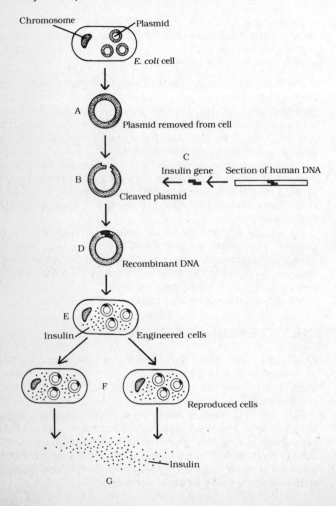

molecule at a particular place but no other. (These enzymes normally operate against *viruses* that prey on bacteria.) Another molecule, *DNA ligase* (Latin *ligare*, to bind), acts like a glue to hold together freshly joined sections of DNA.

Leaving out details, this is the gene-splicing process for human insulin:

1. Plasmids (A) removed from cells of *E. coli* are mixed with a restriction enzyme that cleaves them (B), leaving an open space in each ring.
2. In a separate operation, DNA from human cells is treated with other restriction enzymes; they cut away the desired DNA section (C) — the gene that codes for insulin production — and this is placed in a solution containing the cleaved plasmids from step 1, along with DNA ligase. The human DNA joins the bacterial DNA in the plasmid, producing recombinant DNA (D). More picturesquely, this is sometimes called a *plasmid chimera*, after the chimera of Greek mythology, a monster assembled from the parts of a goat, a lion, and a serpent.
3. The chimeras are placed in a solution containing normal *E. coli* cells, which they enter. Such engineered *E. coli* cells produce tiny amounts of *human* insulin (E). When these cells divide (F), the reproduced cells, and all the successive generations, have the human gene and therefore the ability to produce human insulin (G). In the ideal cultural conditions of the laboratory, a few cells give rise to thousands of millions of identical cells in a few days — the basis of the insulin 'factory'.

The promise Genetic engineering is an enormous scientific leap forward.

- By 1988 hundreds of human genes had been engineered. Some engineered substances: human insulin, INTERFERON for use against viruses and some types of cancer, and a vaccine for one form of hepatitis. Human growth hormone (*somatrem*) was available. Scientists were producing and testing genetically engineered vaccines against malaria and AIDS, as well as engineered products to treat heart attacks, hypertension, and haemophilia.
- Genetically engineered *tissue plasminogen activator* (tPA) has been used with great success to dissolve blood clots in the arteries of the heart, the cause of most heart attacks. Although tPA is produced normally by the body, the amounts needed for treating heart attacks are available only through genetic engineering.

- Successful transfers of genes in laboratory mice suggest the treatment of genetic diseases by implantation of genes. Towards the end of 1988 the first experiment to insert genetically altered cells into humans was approved by the review board of the United States National Institutes of Health.
- Engineered bacteria can be used to introduce new kinds of genes into plants. Plants that produce more nutrients, especially more protein, would help to solve the food problems of developing nations, as would plants able to grow with little or no fertilizer.
- In industry, research under way may provide engineered bacteria that can convert waste wood (the wastes of papermaking, for example) into sugar and into wood alcohol, valuable as an antifreeze, a solvent, and the starting point in the production of many other chemicals.

The problems Early in the development of genetic engineering there were fears that changed organisms might be dangerous: for example, that *E. coli*, our usually harmless colonic tenants, might accidentally be made pathogenic (disease-producing), with epidemic results. Such fears led to unprecedented discussions by hundreds of scientists, with public and governmental participation. Guidelines were designed for safety in research and use of recombinant DNA technology; to date, no serious problems have arisen.

However, many ethical concerns remain. Among them are these:

- Is it acceptable to create new kinds of cows, sheep, and other domestic animals? In 1987 the US Patent Office granted inventors the right to patent such animals, created by genetic engineering and by late 1988 there were moves to grant similar patents to scientists in EEC countries. Going a step further, is it acceptable to use *human* genes to modify animals? For example, using genes for human growth factor, scientists have produced pigs that eat less and are leaner, desirable traits from a commercial point of view. The genes are passed from one generation to the next.
- In 1987 the US Patent Office ruled that genetically changed human beings may not be patented (what a relief!) on the grounds that the antislavery amendment of the US Constitution forbids the ownership of humans. However, since human genes have been routinely and successfully implanted in animals, should not some line be drawn beyond which scientists may not go? Where should that human vs. nonhuman line be? And who should draw it?
- Is it acceptable to use recombinant DNA technology to produce 'improved' (i.e., more deadly) toxins for biological warfare?

- The techniques that may prevent genetic diseases might also be used to produce 'superior' people of 'greater intelligence' or different appearance. Is that acceptable? Who will judge?
    See also DNA AND RNA.

GENETIC FINGERPRINTING   See DNA FINGERPRINTING

GENETICS   See BIOLOGY

GENETIC SCREENING   See AMNIOCENTESIS

GENOME   See GENETIC DISEASES

GEOCHEMISTRY   See CHEMISTRY

GEOPHYSICS   See PHYSICS

GEOSTATIONARY (GEOSYNCHRONOUS) ORBIT   See SPACE TRAVEL

GEOTHERMAL ENERGY

Heat within the earth, mainly the product of radioactivity.
    See also ENERGY RESOURCES.

GeV   See ELECTRON VOLT

GIANT PLANETS   See SOLAR SYSTEM

GIGA-   See ELECTRON VOLT

GIGO

Acronym for 'garbage in, garbage out', a slogan used by computer programmers to remind themselves that computers, no matter how expensive and sophisticated, are not thinking machines — they are only as good as the material with which they are programmed.

GLAND, DUCTLESS   See HORMONE

GLANDULAR FEVER   See HERPESVIRUS DISEASES

GLIDER

An aircraft that is heavier than air, with fixed wings but no self-contained means of propulsion. Gliders are put into motion by launching from the ground by towing or by some other means. They can also be towed aloft by an aeroplane and released.

To stay aloft, the pilot seeks currents of rising warm air (*thermals*) and circles within them. To fly across country, he or she flies from thermal to thermal, gaining altitude within each then sinking gradually towards the next. Obviously a skilled glider pilot is good not only at controlling gliders, but also at detecting evidence of thermals.

Soaring is a sport in which the glider pilot takes advantage of thermals and other air currents to gain altitude (the record is over 14,000 metres, nearly 46,000 feet) or to travel long distances (nearly 1500 kilometres, over 900 miles, is the record).

GLUCAGON See DIABETES

GLUCOCORTICOIDS See STEROIDS

GLUCOSE See DIABETES

GLUINO See GRAND UNIFIED THEORY

GLUON

A particle that is believed to transmit the energy in the strong nuclear force, the main force holding together the atomic nucleus.
See also GRAND UNIFIED THEORY; NUCLEAR ENERGY.

GLYCINE See PROTEINS AND LIFE

GLYCOGEN See DIABETES

GRAFT-VS.-HOST DISEASE See IMMUNE SYSTEM

GRAND UNIFIED THEORY (GUT)

An attempt, almost complete, to describe the four known fundamental forces of the universe with one or two basic sets of mathematical equations. The forces are as follows:

1. *Gravitation* (gravity to us earthlings)
2. *Electromagnetism* (which includes electricity, magnetism, X-rays, light, radio, and several related areas)
3. *The weak nuclear force* (to be desccribed later)
4. *The strong nuclear force* (also described later)

GRAND UNIFIED THEORY (GUT)

The search for the grand unified theory (GUT) is being carried on in laboratories in several nations, employing huge PARTICLE ACCELERATORS, at great expense; yet success, if it comes, is not expected to produce anything 'practical', such as free electricity or antigravity belts. The laudable goal is, through basic research, to achieve an understanding of the fundamental 'stuff' of the universe — its matter and energy.

Let's take a brief look into the four forces, asking three questions: How do they work? What do they do? How strong are they?

Imagine two astronauts drifting free inside a spaceship in orbit, exercising by tossing a basketball back and forth. Each toss pushes the tosser slightly backwards; each catch pushes the catcher slightly backwards. So the *force* exerted by the astronauts' muscles is being transmitted from one astronaut to another, by the exchange, back and forth, of a *substance*, the basketball. If the exchanges were so speedy as to blur the ball into invisibility, you might think there was a 'field of repulsion' between the two astronauts pushing them apart. This analogy holds for all four basic forces we are about to examine. All the forces are transmitted by an exchange of particles — some kinds produce a field of repulsion, and others produce a field of attraction.

We'll begin with the force that is most familiar, because it never lets (you) up: gravity.

Force 1: gravitation The particle that transmits the force of gravitation has not yet been discovered, but scientists are so confident of its existence that a name awaits it: the *graviton*. Newton's apple fell to the ground, say the scientists, and countless apples have fallen since, as the result of a continual exchange of gravitons creating a field of attraction between the earth and the apples. The earth, however, is tremendously massive and hard to budge (it has great inertia). The little lightweight apple is, so to speak, a pushover, with very little inertia. So the massive earth moves the lightweight apple a large amount, while the apple moves the earth (yes!) an infinitesimally tiny amount.

Force 2: electromagnetism Astronauts in space have proved that it's possible to get along without gravitation, but they — and we — would literally come apart without the help of force 2. Everybody and everything — every atom and molecule in the universe — is kept together by the force of electromagnetism. Protons and electrons attract each other, thus forming atoms. Atoms attract each other, forming molecules. Molecules attract each other, forming spiderwebs, kitchen tables, and you, among other things. But attraction is not our only debt to electromagnetism. Protons repel other protons; electrons repel other electrons. This repulsion keeps things separate. You don't sink into a concrete pavement because there's a field of repulsion between the electrons in the top layer of concrete and the electrons in the bottom layer of your shoes (or your bare feet, for that matter). Repulsion, too, helps to keep a stream of electrons moving along in a wire (an electric current) when one end of the stream is pushed by an electric generator or by the chemical action in a battery.

The particle that transmits electromagnetic force is the *photon*. Its discovery is due principally to the research of Max Planck, in 1900, and Albert Einstein, in 1905. Nothing about photons is believable by commonsense rules, but their existence and behaviour have been proved by thousands of experiments that confirm these unexpected traits:

- Photons have no mass — they are weightless.
- Photons vary in energy. Those exchanged between a proton and an electron can be 10^{37} (1 followed by 37 zeros) times stronger than the gravitational attraction between the particles.
- Photons exist only while in motion. Stop them and they cease to be, but their energy is absorbed by the thing that stopped them.
- Some kinds of photons are packets of light energy. Others make up X-rays, radio waves, and other forms of energy transmitted by electromagnetic radiation.

Force 3: the weak nuclear force This force (also called the *weak interaction*) is unfamiliar to most people because its action extends only within the limits of the atomic nucleus, a tiny domain indeed. In many ways it seems to be a relative of the electromagnetic force. The two together have been dubbed the *electroweak force*. There are several weak-force transmitters, including particles called the W−, the W+, and the Z°. Few everyday phenomena are associated with the weak force; an example is one of the 'decay' steps that produce the glow in a radium dial.

Force 4: the strong nuclear force By far the mightiest of the four forces, the strong nuclear force (also called the *strong interaction*) holds the parts of the nucleus of an atom together. It is about 10^{39} times as strong as the gravitational force between those parts. The strong nuclear force holds together the protons of the nucleus, which would otherwise repel each other violently by their electromagnetic force (force 2). Such repulsion is effected under controlled conditions, gradually in nuclear power plants or all of a sudden in one tremendous smash in fission bombs. The holding-together effect of the strong nuclear force, whether in power plants or bombs or in less threatening objects such as this book and you, is exerted in the same way, by particles named *gluons*.

To summarize this too-brief summary of the grand unified theory, this is where the theory stands at present:

1. Gravitation awaits the identification of its particle, the graviton. Gravitation *seems* to bear no relation to the other forces.
2. The electromagnetic force is related to the weak nuclear force — enough to share a group name, the electroweak force.
3. The strong force seems to show a relationship to the electroweak force.
4. Eventually, then, we may expect the physical universe to be describable in terms of two sets of equations:

    a. Those dealing with gravitation
    b. Those dealing with the electroweak-strong force

Some physicists think that eventually all four forces will be described within one super grand unified theory. There are various approaches to this grandeur, all under the general heading of SUSY, an acronym for *su*per*sy*mmetry. The various SUSYs require the existence of 'partners' to the existing known subatomic particles. Some that are being searched for right now have names waiting for them:

gravitino   (partner to the graviton)
photino   (partner to the photon)
selectron (partner to the electron)
gluino   (partner to the gluon)

GRANITE *See* ROCK CLASSIFICATION

GRAPHICS (WITH COMPUTERS)
See COMPUTER GRAPHICS

GRAVITATION

The mutual attraction between all the particles of matter in the universe. When measured locally — for example, at the surface of a planet — it's called gravity, or weight (Latin *gravitas*, weight). But the all-embracing universal term is *gravitation*. The strength of the gravitational force between any two bodies depends on two factors:

1. The amount of matter contained in the two bodies. The more matter, the greater the gravitational force. Buy a ticket to the

moon and have the time of your life, cavorting over the lunar landscape, leaping in lovely 10-metre arcs. You contain the same amount of matter you did back home, but the moon contains only about one-eightieth as much matter as the earth.

2. The gravitational attraction is proportional to the inverse square of the distance between the two bodies. Move twice as far away, and the force is only a quarter as great. Move three times as far, and it's a ninth as great. These distances refer to the centres of gravity of the two objects. Right now you are about 6400 kilometres (roughly 4000 miles) away from the centre of gravity of the earth; you weigh X kilograms (or pounds). Move up into space an additional 6400 kilometres (double the distance) and the gravitational force is only one-quarter as much: you weigh $x/4$ kilograms. At 12,800 kilometres (8000 miles) above the earth (triple the distance), you weigh $x/9$ kilograms.

Suppose you go the other way, all the way down to the centre of the earth. There, surrounded on all sides by the matter of the earth pulling evenly from all directions, you weigh ... nothing! You are weightless because all the gravitational forces are evenly balanced around you.

Although gravitation is the commonest of forces, its method of operation is not known. However, an explanation seems to be waiting in the wings, according to which the force is transmitted by particles called *gravitons*. These have no weight (mass) themselves, travel at the speed of light, and are accompanied by waves called *gravity waves*.

See also GRAND UNIFIED THEORY.

GRAVITATIONAL COLLAPSE   *See* STELLAR EVOLUTION

GRAVITATIONAL FIELD   *See* FIELD THEORY

GRAVITINO   *See* GRAND UNIFIED THEORY

GRAVITON

The name for a particle, as yet undiscovered, that transmits the force of GRAVITATION.

See also GRAND UNIFIED THEORY.

GRAVITY   *See* GRAVITATION

GREENHOUSE EFFECT

Without the greenhouse effect, our atmosphere would be about 18°C (32°F) cooler than it is. The atmosphere behaves somewhat like a greenhouse. Sunlight warms a greenhouse because its rays can pass through transparent plastic and glass. Some of the sunlight is absorbed by the plants, soil, and air in the greenhouse. The rest of the sunlight is reflected. This reflected light (infrared light, also called heat rays) is of a longer wavelength, unable to pass through glass. Instead, it is reflected back into the greenhouse, raising the temperature of everything there. In the case of our huge 'greenhouse' — the earth and its atmosphere — the reflectors are clouds, dust particles, carbon dioxide gas, and water vapour, which 'trap' infrared radiation in the lower layers of the atmosphere.

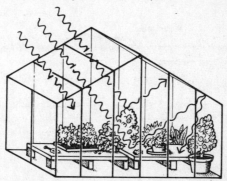

The earth's atmosphere is changing because of human activities, foremost of which is the emission of carbon dioxide when fuel is burned for motor vehicles, industry, and heating. Scientists estimate that the carbon dioxide content of the atmosphere, now 0.035%, will double in less than a century, causing an increase of 3° to 4°C (5° to 7°F) in the earth's temperature. The consequent acceleration in the melting of glaciers and snowfields could raise the level of the oceans by 70 centimetres (more than 2 feet) per century, to begin the drowning of most of the earth's seaports.

GROUND STATE (OF ELECTRONS) See ENERGY

GROUNDWATER

A better name for it would be *under*groundwater. Water from rain and melted snow soaks the ground, then seeps down through the soil and rock. Pores in some kinds of rock and cracks in others make them permeable to water. The water continues down until it reaches a point where all the pores have been saturated or where the rock is not porous (impermeable). Such water-bearing regions are called aquifers (Latin *aqui*, water + *fer*, to bear). Here bodies of groundwater collect. The top level of the groundwater is the *water table*, or *groundwater level*, and its depth changes constantly, rising when there is precipitation and falling during dry periods.

Wells are dug down to below the groundwater level, and the water is pumped up for use. Water shortages have become a common problem in places where excessive groundwater pumping occurs, which results in prolonged or permanent lowering of the water table.

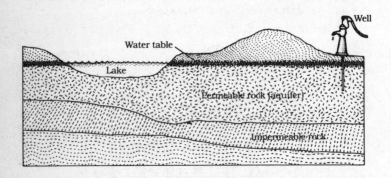

GROWTH HORMONE   See HORMONE; GENETIC ENGINEERING

GUANINE   See DNA AND RNA

GUARD CELLS   See TRANSPIRATION

GUT   See GRAND UNIFIED THEORY

HAEMODIALYSIS See KIDNEY DIALYSIS

HAEMOGLOBIN See GENETIC DISEASES

HAEMOPHILIA

An inherited disease, occurring almost entirely in males, in which the clotting ability of the blood is severely impaired. Normally, blood flowing from an injury sets up a chain of complex chemical reactions that ends in the formation of a blood clot, a damlike tangle of threads of *fibrin*. In haemophiliacs, however, one of the numerous substances involved in the reactions, *factor VIII*, is missing or in short supply. The chain of reactions is affected, and clot formation is slowed or stopped. Severely affected haemophiliacs must carefully avoid activities with potential for injury. Bleeding is treated with concentrates of factor VIII prepared from human blood plasma. Blood transfusions may be needed for tooth extractions and minor surgery. Towards the end of 1988 a synthetic blood clotting factor was successfully used in treatment of bleeding for the first time.

HALF-LIFE See UNCERTAINTY PRINCIPLE

HALLEY'S COMET See SOLAR SYSTEM

HARD DISK See FLOPPY DISK

HARDWARE

The machinery part of a computer, consisting of mechanical, magnetic, electrical, and electronic parts. The hardware awaits instructions from the *software*, which contains a program in the form of tapes, disks, punched cards, or human fingers (directed, of course, by human brains).

HDL See CHOLESTEROL

HEART See CIRCULATORY SYSTEM

HEART ATTACK

A failure of the pumping action of the heart and the greatest single cause of death in most developed countries, including both Britain and the United States. Scotland, with about 18,000 deaths per year, has the highest death rate (more than 1 in 300) from heart disease of any country in the world.

The heart pumps blood through blood vessels (arteries) that branch to reach every part of the body (see CIRCULATORY SYSTEM). The blood delivers its load of nourishing materials to the body cells and, through veins, takes away the waste materials they produce. Consider the staggering job done by the heart. It is composed largely of *myocardium* (Greek, *myo*, muscle + *kardia*, heart) — thick, powerful muscle tissue that contracts and relaxes some 70 times per minute. Except for the intervals between heartbeats (less than ½ second), it works for a lifetime without rest, pumping about 30 teaspoons of blood with each beat. That adds up to some 680 litres (150 gallons) per hour. Clearly, the hardworking heart needs a generous supply of blood for itself. It is, in fact, fed by a special set of blood vessels, the *coronary arteries*. These vessels sit like a crown over the outside of the heart (Latin *corona*, crown), and their branches extend deep into the myocardium.

Most often a heart attack results from the closing (occlusion) of a coronary artery by a blood clot or by plaques of fatty material (see ARTERIOSCLEROSIS). Sometimes a spasm in an atery may close it down. Other factors that may be involved are smoking, hypertension, and diabetes. In any case the myocardium recieves less blood than it needs. An acute shortage causes the death of some of the tissue, and the heart's pumping act is impaired. The dead, clogged area is called a *myocardial infarct or infarction* (Latin, *infarcire*, 'to stuff in').

HEART-LUNG MACHINE See OPEN-HEART SURGERY

HEART SURGERY See OPEN-HEART SURGERY

HEART-TRANSPLANT See OPEN-HEART SURGERY

HEAT ENERGY See ENERGY

HEAT LIGHTNING See LIGHTNING AND THUNDER

HEAT RAYS See ENERGY

HEAT SHIELD See SPACE TRAVEL

HELICOPTER

An aircraft that is heavier than air, is self-propelled, and has engine-driven rotating wings.
See SPACE TRAVEL.

HELIUM NUCLEI   *See* ALPHA PARTICLE

HELPER CELLS   *See* IMMUNE SYSTEM

HERPESVIRUS DISEASES

A family of VIRUSES that cause several diseases characterized by spreading sores or blisters (Greek *herpein*, to creep).

Herpes simplex I   Causes cold sores on the mouth and face.

Herpes simplex II   Causes painful sores on the mucous membranes (linings) of the genital organs. The disease is spread by sexual contact and was rampant in the early 1980s, reaching epidemic levels in the United States with an estimated 20 million victims. A newly developed drug, *acyclovir*, is useful in alleviating the pain of the recurrent attacks, but no cure is known. There is hope that a newly developed vaccine, produced by GENETIC ENGINEERING, will offer protection from the virus. Although the disease is not life-threatening in adults, there is high risk of infection to children born to mothers in the active phase of the disease. In such cases, caesarean section reduces the risk to the infant.

Herpes zoster   Causes *chicken pox*, usually in children; it may reappear later in life, causing the painful nervous system condition known as *shingles*.

Epstein-Barr virus (EBV)   Causes *infectious mononucleosis* (also known as *glandular fever*), a relatively mild disease of young adults; it is also suspected of causing *Burkitt's lymphoma*, a type of cancer found in Central Africa.

HERTZ   *See* RADIO

HIGH BLOOD PRESSURE   *See* HYPERTENSION

HIGH-DENSITY LIPOPROTEIN (HDL) CHOLESTEROL   *See* CHOLESTEROL

HIGH-ENERGY PHYSICS   *See* PHYSICS

HIGH FIDELITY

Also called *hi-fi*. A measure of the accuracy with which a system (particularly a sound-reproducing system) delivers the message that has been put into it. In a record-playing system, for example, the message (music or speech) fixed in the wiggly grooves of a record goes through a series of transformations before you can hear it. The first in line to receive the message is the stylus (needle), which is made to vibrate by the curving, rotating grooves. The vibrations are converted into electric currents by a cartridge (pickup), and the currents are then fed into a series of electronic circuits to be amplified and sent into speakers that convert the electrical signals into sound waves.

At each step the message is altered by a less-than-perfect (lower-fidelity) transmission. These deviations add up from step to step. The main kind of deviation is measured in units called *total harmonic distortion*, or *THD*. The lower the THD, the higher the fidelity. A cheap record player, for example, has a THD of several per cent, while a good amplifier may have a THD of 0.005% or less.

See also SOUND RECORDING.

HIGH-LEVEL LANGUAGE   *See* COMPUTER LANGUAGES

HISTAMINE

An amino acid derivative normally secreted by the body in very small amounts. The amount increases when activity in the IMMUNE SYSTEM increases, as in hay fever or other allergic reactions. The extra histamine causes sneezing, itching, running eyes, hives, and other distressing symptoms. A number of drugs — *antihistamines* — can be used to counteract the histamine. Injury to a body organ releases considerable amounts of histamine, which causes blood vessels to dilate, thereby lowering the blood pressure.

HISTIDINE   *See* DNA AND RNA

HISTOLOGY   *See* BIOLOGY

HIV   *See* AIDS

HOLOGRAPHY

A method of making three-dimensional images that change as you change your viewing position. Move from front to side, for example, and the front view of a face turns gradually to a profile. You see a whole picture, or *hologram* (Greek *holos*, complete). At present only still pictures are possible, but technology is being perfected for motion. With a holographic cassette in a holographic projector, you may someday be able to watch a play, with the characters speaking and moving in front of you not on a screen or TV set but in the actual space of your living room! (But don't try to prevent the villain from menacing the heroine. He's only a light picture, and you can walk right through him.)

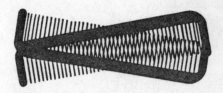

Holography is based on the principle of *interference patterns*. To make an interference pattern, hold two matching combs flat against each other, at a slight angle, and observe them against the light. You will see an interference pattern of geometric shapes that change as you alter the angle between the combs.

Now for the interference patterns that produce holographic images. A beam of laser light is split into two matched beams (like two combs). One of the beams, which we'll call A, is reflected by a beam splitter onto lens A, which focuses it onto mirror A. It is reflected by the mirror onto the subject (a person, for example) and then onto an unexposed film. The second beam (call it B) passes through the beam splitter to mirror B. The mirror reflects it onto lens B, which focuses it onto the same unexposed film. The light waves in the two beams, like the teeth in the combs, don't match exactly. They produce many complex interference patterns on the film. Then the film is developed, placed in a holographic projector, and illuminated by a laser beam. The interference patterns on the film break the smooth, straight laser beam into tiny points of light and shadow that reconstruct the form of the original subject. These points don't require a screen to shine on — they form images in the air. Therefore, they can make rounded, three-dimensional shapes rather than flat two-dimensional ones.

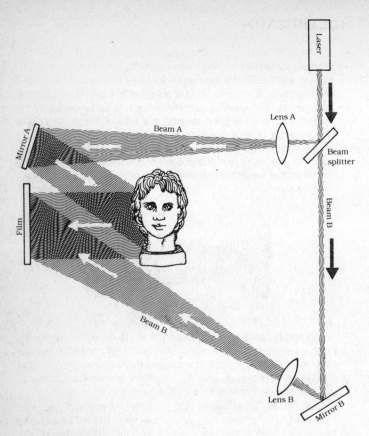

HORMONE

A substance that is sometimes called a 'chemical messenger' because its effects may occur at some distance from the gland in which it is secreted. In such a ductless, or *endocrine*, gland, the hormone (Greek *horman*, to set in motion) passes directly into the blood that circulates through the gland. The blood carries the hormone throughout the body until it reaches the target cells whose functions it regulates. For example, the *thyroid* gland in the neck secretes two related hormones, *thyroxine* and *triiodothyronine*, which regulate the rate of oxygen consumption by the body. In this case, the target cells are all the cells of the body.

The workings of the endocrine glands and of the nervous system are intimately linked. Most endocrines secrete more than one hormone and control more than one function. Here are some examples of hormones and their work:

- *Parathyroid hormone* regulates the level of calcium and phosphorus in the blood; it is secreted by four tiny glands, the *parathyroids*, mounted on the thyroid gland.
- *Adrenaline* and *noradrenaline*, which regulate blood pressure, among other jobs, are secreted by the *adrenal glands*, located at the top of the kidneys.
- *Insulin* controls the storage and use of sugar, and its concentration in the blood; it is secreted by groups of cells in the pancreas known as the *islets of Langerhans*.
- *Growth hormone*, which promotes the growth of bones and muscles, is secreted by the *pituitary*, a gland attached to the

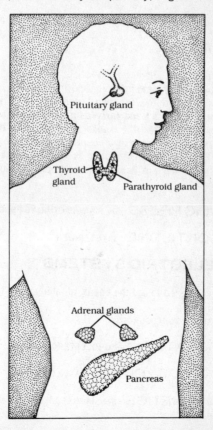

lower surface of the brain. It secretes a large number of other hormones, some of which regulate the operation of other endocrine glands.
See also STEROIDS.

HOT-DRY-ROCK TECHNOLOGY
See ENERGY RESOURCES

HUMAN GROWTH HORMONE
See HORMONE; GENETIC ENGINEERING

HUMAN IMMUNODEFICIENCY VIRUS   *See* AIDS

HUMAN INSULIN   *See* GENETIC ENGINEERING

HUMIDITY   *See* RELATIVE HUMIDITY

HURRICANE

A huge (up to 800 kilometres, or about 500 miles, in diameter) circular whirlwind of great intensity, moving from about 120 to over 320 kilometres (75 to 200 miles) per hour. It is energized by inwards-moving cold air and upwards-moving warm air around the relatively calm centre (the eye). Hurricanes in the Northern Hemisphere spiral inwards in a counterclockwise direction; in the Southern Hemisphere their motion is clockwise. Also known as typhoons in the western Pacific and as tropical cyclones over the Indian Ocean.

HYDRAULIC PRESS   *See* MANUFACTURING PROCESSES

HYDROCORTISONE   *See* STEROIDS

HYDROELECTRIC SYSTEMS

Machinery designed to use the energy of falling water for generating electricity.
See also ENERGY RESOURCES.

HYDROGENATION   *See* CHOLESTEROL

HYDROLOGIC CYCLE   *See* WATER CYCLE

HYDROMAGNETICS   *See* MAGNETOHYDRODYNAMICS

HYPERON See ELEMENTARY PARTICLES

HYPERTENSION

Abnormally high BLOOD PRESSURE. It is estimated that 10 to 15% of the population have some degree of hypertension at any one time. Because there are no symptoms, many do not know they are affected.

The cause of hypertension is not known. There is no known cure, but a variety of techniques are used successfully to lower the blood pressure. Among them are low-salt diets, weight-loss diets and exercise for the overweight, and a variety of drugs. Hypertensive people should stop smoking, use alcohol only in moderation, and avoid stress as much as possible. Neglected hypertension is dangerous because it can lead to kidney failure, retinal disorders, HEART ATTACKS, and STROKE.

Blood pressure is expressed in simple numbers — for example, 120/80 — but the interpretation of these numbers for a given individual requires great skill. A reading that is normal for one person may be high or low for another. A diagnosis of hypertension should be made only by a doctor.

HYPOTHESIS See SCIENTIFIC TERMS

HZ See RADIO

IGNEOUS ROCK See ROCK CLASSIFICATION

IGY See INTERNATIONAL GEOPHYSICAL YEAR

ILLUMINATION NOISE See NOISE

IMMUNE RESPONSE See IMMUNE SYSTEM

IMMUNE SYSTEM

The body's Ministry of Defence. The strong temptation to talk about it in military terms will not be resisted.

Throughout life, we are besieged by foreign forces — viruses, bacteria, fungi, plant pollens, insect and animal poisons, and many other strangers whose effects on us range from imperceptible to fatal. We are protected by three main lines of defence:

- Defence 1 comprises the skin, hairs in the nose, mucus in the nose and throat, and other such simple but generally effective mechanical barriers.
- Defence 2 is a standing army of white blood cells. These are carried by the bloodstream to all parts of the body, but they can also crawl freely among the body tissues. They engulf and devour bacteria and other invaders — and some are themselves killed in the process. They are closely followed by *macrophages* (Greek *macros*, large + *phagos*, eater) to clean away the debris. The pus seen in an infected area — in a pimple or an abscess, for example — is a mixture of dead invaders, defenders, and tissue cells in blood fluid (serum).
- Defence 3 is a reserve army of antibodies. They do battle against foreign substances called antigens, which may cause damage by getting involved in the body's normal chemical activities. The immune system creates these antibodies, and each is specific: it works against the antigen that triggered its creation, or a closely related one, but not against others. (For example, typhoid antibodies don't work against smallpox.) The antibody and antigen combine, which action inactivates the antigen. Think of the antigen as a set of sharp fingernails whose scratch power is overcome by a made-to-order glove (the antibody) fitting over them.

In 1796 Edward Jenner (1749–1823), an English physician, tested the belief that people who had recovered from cowpox, a relatively mild disease, were unlikely to contract smallpox, a frightful disease that disfigured or killed enormous numbers of people. He inoculated a young boy with matter (pus) from a cowpox sore. After the child had come down with cowpox and recovered, Jenner inoculated him with smallpox matter. The boy remained healthy; he had become immune by building up antibodies against the cowpox virus and the closely related smallpox virus. Today, thanks to Jenner's pioneering, smallpox seems to have been eradicated from the earth. Better still, the principle of vaccination (Latin *vaccinus*, of cows) has been turned successfully against a long list of killing and crippling diseases, including polio, rabies, cholera, plague, diphtheria, and tetanus.

The *immune response* — the production of antibodies by the immune system — is extremely complex and not fully understood, but let's touch on a few major facts, skipping many lesser details.

The main source of cells for making antibodies is the bone marrow, which produces *lymphocytes* (Latin, watery cells). Some lymphocytes develop into *B cells* (B is for *bone*). A B cell exposed to an antigen produces a small amount of *immunoglobulin* — molecules of antibody. It also 'remembers' by retaining the chemical configuration, or pattern, of the antigen. On later exposures to antigens of the same kind, the B cell multiplies rapidly, resulting in a large number of identical cells — clones — that can make the antibody. The antigen extermination is done mainly against viruses and bacteria circulating in the blood.

Other lymphocytes, the *T cells (T* is for *thymus)*, become active only after passing through the thymus, a gland located in the chest cavity, at the base of the neck. T cells work mainly within the body cells, against viruses, bacteria, and, interestingly, cancer cells. T cells come in at least two varieties: *helper cells*, which appear to stimulate activity in the immune system, and *suppressor cells*, which slow it down as the invaders are overcome. Imbalances between the two kinds of cells may be responsible for some faults in the immune system. For example, AIDS is a disease in which the immune system has effectively stopped working because the balance is tipped in favour of the suppressor cells.

T cells are responsible for rejection of transplanted (therefore foreign) skin, kidneys, and other organs; such rejection is called *graft-vs.-host disease*. The rejection can be prevented by the use of immunosuppressant drugs, notably cyclosporin. Such drugs must be used with great care, because they also suppress the body's ability to fight infectious diseases.

The T cells also produce INTERFERON, a powerful antiviral substance.

Although the immune system is astonishingly good at its job, there are occasional lapses:

Excess of zeal The immune system fights antigens of all kinds, including harmless ones like pollens, animal fur, and shellfish; the body is said to be *sensitized* to these. In some hypersensitive people, a second or third exposure to a particular antigen may provoke an overreaction: as antigen and antibody combine, the cells release a substance called HISTAMINE. It causes the itching, sneezing, weeping eyes, and other distressing symptoms familiar to *allergy* sufferers.

Anaphylaxis is an extreme and dangerous form of allergic reaction. The victim experiences faintness, great difficulty in breathing, heart palpitations, and a drop in blood pressure. The anaphylactic reaction may follow injection of an antigen, as with a bee sting or an antibiotic, or eating or inhaling a substance containing the antigen.

Confused identities The immune system sometimes fails to recognize some of the body's own materials; it reacts to them as if they were foreign antigens. Thus it produces antibodies that attack the body itself. Some scientists suggest that rheumatoid arthritis, multiple sclerosis, and juvenile-onset DIABETES are the result of this *autoimmunity*.

Failure to operate Occasionally, children are born with a condition in which their immune systems are unable to form any antibodies. Such infants, defenceless against infectious diseases, rarely survive more than a year. Some of them have lived longer by being kept in bubblelike sterile tents. Recently, the use of bone-marrow transplants has given some of these children a measure of immunity, allowing them to live outside the bubble.

IMMUNOGLOBULINS   See IMMUNE SYSTEM

IMMUNOSUPPRESSANTS   See IMMUNE SYSTEM

INBORN ERRORS OF METABOLISM
See GENETIC DISEASES

INDUCED CURRENTS   See TRANSFORMER

INDUCTANCE   See CROSS TALK

INDUCTION   See SCIENTIFIC TERMS

INDUSTRIAL CHEMISTRY   See CHEMISTRY

INERTIA *See* MASS

INERTIAL MASS *See* RELATIVITY

INFECTIOUS MONONUCLEOSIS
See HERPESVIRUS DISEASES

INFILTRATION (OF CANCER) *See* CANCER

INFRARED RAYS *See* RADIANT ENERGY

INGOTS (STEEL) *See* MANUFACTURING PROCESSES

INJECTION MOULDING *See* MANUFACTURING PROCESSES

INORGANIC CHEMISTRY *See* CHEMISTRY

INORGANIC VS. ORGANIC *See* PROTEINS AND LIFE

INPUT *See* COMPUTER

INSULIN *See* DIABETES

INTEGRAL CALCULUS *See* CALCULUS

INTEGRATED CIRCUIT *See* CHIP

INTERFACE

The meeting place between two different systems. A French-English interpreter is a human interface between English-speaking diplomats and French-speaking diplomats. An electronic interface inside a pocket calculator converts the ordinary Arabic digits 1, 2, 3, and so on, which are not usable by binary circuits, into binary digits 0 and 1. After the binary answer has been obtained, it is presented to another interface in the calculator, which converts the answer back into an Arabic-numeral display.

A *modem* (short for *m*odulator-*dem*odulator) is an interface that enables two computers to communicate via telephone lines. The sender's modem modulates (converts) the digital language of a computer (electrical pulses and spaces) into analogue language, like the musical beeps of a push-button telephone. The beeps, thousands per second, travel over the phone wires to the receiver's modem, which demodulates the tones back into digital blips that the computer can use.

See also ASCII.

INTERFERENCE PATTERNS See HOLOGRAPHY

INTERFEROMETRY

A system of precision measurement using waves — radio waves, light waves, or sound waves — as measuring sticks. Interferometers are used in taking measurements as large as the distance between two stars and as small as the difference in thickness between two human hairs of different colours (blond hairs are usually thinner than black).

You can illustrate the working principle of an interferometer by holding both hands, fingers outstretched, against the sky. You can see light areas (sky) and dark areas (your fingers). Then place one hand over the other, palms together, so that the fingers of the right hand fit in the spaces of the left. Now you can't see light, because the fingers of one hand *interfere* with the light in the spaces of the other hand.

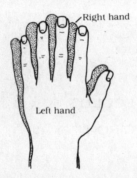

Next, still looking at the sky, slide one hand to the left, for the width of one finger. At the moment of sliding over, you will see a flash of light, then darkness as the fingers of one hand again

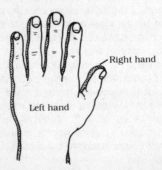

interfere with the light in the spaces of the other hand. Try again, sliding across two fingers. Two flashes. Three fingers, three flashes. If you knew the width of your fingers (assuming them to be all the same width), you would know how far you had moved them — not by measuring but by counting flashes and interferences.

However, a finger's thickness is a rather coarse unit of measurement. Try two identical combs, where the unit of measure is one tooth. If, for example, there are ten teeth per centimetre, each flash would indicate a movement of $1/10$ centimetre. Here's a pair of combs being used to measure the diameter of a penny — the finer the teeth, the finer the measurements. But there's a limit to the fineness of even a fine-toothed comb.

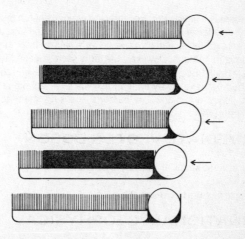

The limit on light waves is much higher. For instance, in light from a helium-neon laser, there are 63,280 waves in the space of one centimetre. Suppose we have two beams of helium-neon laser light as our two 'combs'. When the waves are matched, wave over wave in an interferometer, we see brightness. When we slide one of the beams, the waves cancel each other; dark lines are produced by interference. Light and dark, light and dark, each flash indicates 1/63,280 centimetre, or 0.0000158 cm. That's about how much a railway track bends when a sparrow alights on it. And an interferometer can measure it!

INTERFERON

The story of this remarkable natural substance came close to being 'too little, too late'. It was discovered in 1957 by a team of scientists led by Alick Isaacs, a British virologist, and Jean Lindenmann, a Swiss microbiologist.

There are at least 25 types of interferons, but they are usually discussed in the singular, and we shall do the same.

Interferon is a protein made by certain body cells in response to infection by a VIRUS. Interferon travels in the blood through the body and interferes with a virus's ability to infect healthy cells. It protects against the infecting virus and against many other kinds of viruses as well — it is nonspecific. This gives interferon enormous potential for preventing or overcoming colds, herpes, hepatitis, rabies, and other viral diseases, just as antibiotics triumphed over diseases caused by bacteria. Even more, interferon is effective against certain kinds of cancer.

But there is — or was until recently — a major research problem: the scarcity of interferon. Human cells make only tiny amounts of it, barely enough for scientists to work with, and animal interferon is ineffective in humans. However, the story seems headed for a happy ending. By 1985, interferon was being made in commercial quantities by GENETIC ENGINEERING. This is a technique in which human genes are placed in bacteria, causing them to produce human interferon.

INTERMEDIATE VECTOR BOSON

One of several kinds of particle that transmit energy in the WEAK NUCLEAR FORCE.

See also ELEMENTARY PARTICLES.

INTERNATIONAL GEOPHYSICAL YEAR (IGY)

An unprecedented research effort, sponsored by the International Council of Scientific Unions, in which thousands of scientists from 67 nations cooperated in a study of the earth and its surroundings. The 'year' actually lasted from July 1957 to December 1958, but many of the participants continued their IGY projects beyond its formal end.

Among the many research subjects were the earth's crust, shape, atmosphere, weather, oceans, and polar regions; the sun also came in for its share of study. The earliest artificial satellites of the United States and the Soviet Union were launched in connection with the IGY programme. The first American satellite, *Explorer I*, revealed the existence of the VAN ALLEN BELTS, zones of radiation around the earth.

IGY research was also involved in the discovery that a chain of mainly submerged mid-ocean ridges, some 64,000 kilometres, or 40,000 sinuous miles, long, makes up the world's largest mountain range. And not least, there was the discovery that scientists from countries with diverse political and economic systems could work together productively and harmoniously.

INTERSTELLAR MEDIUM See STELLAR EVOLUTION

INTIMA See ARTERIOSCLEROSIS

INTRUSIVE FORMATION (GEOLOGY)
See ROCK CLASSIFICATION

INVERSE SQUARE LAW

If the light on this page is too dim, you can improve matters by moving closer to the source of light, thus capitalizing on the inverse square law. The law states that energy from a point source (call the bulb a point), if unhindered by mirrors, lenses, or other impediments, spreads out equally in all directions and that its intensity diminishes as the inverse square of the distance. At a distance of 1 metre from the bulb, the strength of the light is, let's say, x. At 2 metres the strength is $\frac{1}{4}x$. At 3 metres, it is $\frac{1}{9}x$. So when you move the book from 3 metres to 2 metres, you have more than doubled the brightness of the light falling on the page.

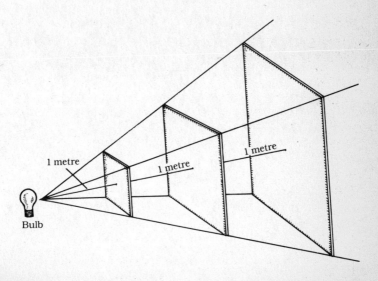

The law of inverse squares applies to any form of energy that spreads out equally in all directions. Magnetism, electromagnetism, and gravity operate under that law. Radio waves and television waves (which are electromagnetic) are strongest at the point of emission at the broadcasting station and weaken rapidly thereafter. A magnet's pull is strongest close to its poles, and so on. The moon's gravity causes the waters of the earth to heap up in tides, while Jupiter, about 26,000 times more massive than the moon, has almost no effect on tides because of its distance from the earth.

By an inverse use of the inverse square law, we can determine distance. For example, an astronomer knows that star A is actually four times as bright as star B (he has ways). Yet both stars measure equally bright in his telescope. Thus he knows that star A is twice as far away.

IN VITRO *See* TISSUE CULTURE

IN VIVO *See* TISSUE CULTURE

IRREGULAR GALAXY *See* GALAXY

ISLETS OF LANGERHANS *See* DIABETES

ISOTOPE *See* NUCLEAR ENERGY

JET

A stream of gas or liquid; frequently used to mean a jet engine or an aeroplane propelled by jet engines.
See also AERODYNAMICS.

K AND k

Both *K* and *k* stand for *kilo-* (Greek *chilioi*, thousand), a multiplying prefix. In the metric system, *k* indicates 1000; thus *kilometre* means 1000 metres. In computer terminology, *K* (for *kilobyte*) refers to memory capacity and is not precisely 1000 but the nearest multiple of 2. For example, a 1K memory has a storage capacity not of 1000 but of 1024 bytes because 2^{10} (2 times itself 10 times) equals 1024. The capacity of a 2K memory is 2048 bytes (1024 times 2). Why this oddity? Because computer memories — like all computer functions — are based on the BINARY NUMBER SYSTEM.

KAON   *See* ELEMENTARY PARTICLES

KAPOSI'S SARCOMA   *See* AIDS

KELVIN   *See* ABSOLUTE ZERO

KEROGEN   *See* SYNTHETIC FUELS

KETONE   *See* DIABETES

KIDNEY DIALYSIS

An artificial method of removing waste products from the blood, also called *haemodialysis*. Normally, this cleansing is performed by the kidneys, but that process is impaired in some kinds of kidney disease. Early dialysis machines were large, cumbersome, and very expensive and could be operated only by health professionals in hospitals. Now there are smaller machines that can be used at home by some patients with non-professional help.

In the process of dialysis (Greek *dialyein*, to break apart or separate), the machine draws blood from the patient, pumps it through membranes of specially designed cellophane, and returns it to the body. This artificial circulation usually continues for several hours. The molecules of dissolved wastes pass through the microscopic pores of the cellophane, but the much larger molecules of the blood, and the blood cells, are held back, thus achieving the separation of wastes from the blood.

The dialysis principle is used commercially for purifying various solutions. A variation, *electrodialysis*, is one of the methods used to desalt water on a large scale.

KILOWATT-HOUR See ELECTRICAL UNITS

KNOWLEDGE-BASED SYSTEM See EXPERT SYSTEM

KWH See ELECTRICAL UNITS

LACCOLITH See ROCK CLASSIFICATION

LAMBDA PARTICLE See ELEMENTARY PARTICLES

LANDING GEAR

The undercarriage of an aeroplane. It usually comprises wheels, brakes, and shock absorbers, and in most aeroplanes it can be withdrawn into the body or wings during flight. Some aeroplanes are fitted with floats or skis for use on water or snow.

See also AERODYNAMICS.

LASER

Acronym for *l*ight *a*mplification by *s*timulated *e*mission of *r*adiation. An appropriate slogan for the laser might be 'In union there is strength'. A laser is an electric apparatus for producing unified light waves that can be exactly controlled, precisely focused, and, when desired, made extremely powerful. Laser light can be shaped to such a narrow, pointed beam that it will burn out the centre of the full stop at the end of this sentence and leave a tiny black ring. It can be aimed precisely enough to destroy a dangerous skin tumour without affecting healthy skin tissue a hundredth of an inch away. Lasers can be powerful enough to cut through solid steel as thick as your fist.

What's different and special about laser light? Imagine this situation: it's necessary to break down a large, heavy wooden door. Available for the job are several people, each with a wooden club. They flail away at the door, each at his own rate and strength – and produce a lot of clatter. This is an incoherent (from the Latin for 'not together') attack on the problem. Someone suggests an improvement: fasten all the clubs into one large club, a battering ram. Assemble all the people around it so that they can run at the door as a team, with their united strength. This is a coherent attack. Success.

The light you live by – sunlight, electric light, the light from a candle or paraffin lamp – is *incoherent*. It's a jumble of different wavelengths and brightnesses, in what seems to be a steady light emitted in every direction. Shaping this jumble into a straight beam, in a torch or car headlight, requires a curved mirror behind the light source and sometimes a lens in front. Still more precise shaping, as in a film projector, requires a curved mirror and several lenses. Even with all these helpers we can't achieve a precise, coherent, pinpoint, concentrated light beam, because the original light source is itself an incoherent mixture. To produce a coherent beam, then, the original light has to be coherent, and that's what a laser is for.

A laser beam produces *coherent light* in two steps. First, it creates packets of light — photons (*see* RADIANT ENERGY) — all having the same wavelength but not yet coherent. (Equal-sized clubs are handed out.) Second, the photons are lined up and organized so that their waves match, crest to crest, trough to trough, all parallel, all coherent. (The clubs are tied together into a single battering ram.)

1. Same wavelength but not yet coherent

2. Coherent

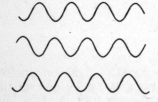
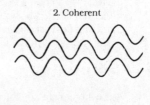

How are these steps accomplished? If you want to go a bit deeper, we must back up slightly.

Ordinary light sources (sun, electric bulb, flame, etc.) don't make photons of the same wavelength; that's because the sources of their energy — hot molecules vibrating randomly — don't move together at exactly the same rate in the same direction at every moment.

Laser light sources are different, however. They don't depend on the random motion of heated molecules. They originate instead in the precise motion of electrons moving from one exact orbit to another. These orbits are determined not by temperature but by the atomic structure of the laser material itself. Each kind of orbital change produces its own special wavelength, its own unique colour. In chromium atoms, for example, a deep red light is generated. Chromium atoms are present in rubies. Artificial rubies (less expensive than natural ones) are used in ruby lasers. Many other kinds of atoms — in solids, liquids, and gases — are used in various lasers, each producing photons with specific wavelengths.

However, even though these photons have the same specific wavelength, they are not yet lined up, matched, coherent. This last step is accomplished (usually) by reflecting photons back and forth between two mirrors, precisely spaced to encourage fitting together (*see* RESONANCE).

A laser is really a kind of basic tool, in the sense that a wedge is a basic tool. A simple wedge can be used to split wood. Make it long and flat and slender and it becomes a knife or sword. Line up a row of wedges and you have a saw. All these uses, and many others, are applications of the wedge principle: a force applied to a broad area is multiplied when transmitted to a point. Similarly, the laser principle — the generation of coherent, precisely direct beams of light energy — has been applied in many ways.

Lasers in action make use of one or both of the following virtues of the laser principle:

Concentration of energy A wide beam of laser light can be focused to an extremely fine point, thus producing a very high temperature at that point (a kind of laser wedge). You might call such a tool a heat knife. Such 'knives' are used by surgeons to produce self-cauterizing cuts. In clothing factories, computer-guided lasers move across dozens of layers of cloth at a time, cutting material for dozens of suits in a minute. In machine shops, lasers cut through steel much faster than saws or other wedge tools. Eye surgeons can 'spot-weld' a detached retina with a low-power laser beam shining right through the transparent lens of the eye. Laser beams have also been used to 'weld' damaged blood vessels without the formation of scar tissue as might happen when stitches are used. In a car factory, high-power laser beams spot-weld the parts of a car body together. Still in the research stage: the nuclear energy fusion process requires a starting temperature of millions of degrees, obtained by concentrated laser beams.

Control Laser light is highly controllable, both in its direction and on-and-offness. Let's look at examples of such applications.

In surveying, a low-power laser beam is aimed across a space (a river, for example) to a mirror that reflects it back into an instrument that clocks the round trip. A computer converts the time measurement into distance (light travels at almost 300,000 kilometres per second, or 186,000 miles per second). For example, if a laser beam travels across a river and back in $1/300,000$ second, the river is 0.5 kilometre (about $1/3$ mile) wide. A mirror on the moon (placed by an Apollo astronaut) is used for periodic checks on the varying earth-moon distance.

Holograms are three-dimensional images in space, like a slide projector image formed in midair, without a screen. This requires highly precise beams of light that can intersect to form points of light — just what a laser apparatus can do so well.

In a telephone system employing FIBRE OPTICS, voice vibrations are converted into pulses of laser light, thousands per second. This would be impossible with ordinary light sources such as tungsten bulbs, which require start-up and cool-down time for each light pulse. Laser light is instantaneous, so millions of pulses can be transmitted in a second. A hair-thin glass fibre can carry several thousand telephone messages at once.

See also HOLOGRAPHY.

LASER PRINTER *See* PRINTER

LASER RECORDING *See* SOUND RECORDING

LATITUDE *See* CELESTIAL COORDINATES

LAVA *See* ROCK CLASSIFICATION

LAW, SCIENTIFIC *See* SCIENTIFIC TERMS

LAW OF INVERSE SQUARES *See* INVERSE SQUARE LAW

LCD AND LED

The two principal methods of forming numbers and letters on instruments such as calculators and digital watches. A basic pattern of seven bars is used to form the digits 0 to 9 and several letters. To form other letters and symbols, more than seven bars are required.

In the LED (*l*ight-emitting *d*iode), the bars are made of a substance that permits an electric current to flow through in one direction only. A substance used in this way is called a *diode*. As the current flows, the diode gives off red, blue, yellow, or other coloured light, depending on the compound of which it is made. For example, gallium phosphide (GaP) emits a green glow. Electric circuits in the instrument selectively turn on the current to the bars to form the various numbers and letters.

In the LCD (*l*iquid crystal *d*isplay), the bars are made of liquid crystals. These are a kind of hybrid material, not quite a liquid and not quite a solid. They can't be poured readily, as with liquids, nor are their molecules locked in place, as with true solids. But the molecules can be rotated slightly by an electric current. When no

```
0123456789
ACEFHIJLPU
```

current flows, the bars are not noticeable, because they reflect light to the same extent as the rest of the display surface. But when a current flows through a bar, its molecules rotate and its ability to reflect light is reduced. That bar appears darker than the area around it and forms part of a number or letter.

You can produce a similar darkening effect, called polarization, with Polaroid sunglasses. Hold the glasses several centimetres from one eye and look through one lens at a shiny, sunlit surface. Rotate the lens and observe the darkening.

Liquid crystals can be made to order to do a particular job. For example, one kind of crystal is sensitive to slight temperature changes. It is used in thermometers where the number representing the temperature appears, then disappears, to be succeeded by a higher or lower number as the temperature changes.

LDL   See CHOLESTEROL

L-DOPA   See PARKINSON'S DISEASE

LEADING EDGE

The front edge of an aircraft wing or of a propeller. The rear edge, reasonably enough, is called the trailing edge.
See also AERODYNAMICS.

LED   See LCD AND LED

LEGIONNAIRE'S DISEASE

Disease discovered in 1976 when nearly 200 members of the American Legion, at a convention in Philadelphia, were mysteriously stricken and 29 died.

The disease begins with chills, fever, cough, and diarrhoea, leading after a few days to pneumonia. The cause of the disease, *Legionella pneumophila,* a previously unknown bacterium, was found by researchers at the US Centers for Disease Control. There is little or no transmission from one person to another; rather, there are clusters of cases, often connected with a particular building – a hotel or hospital, for example, where the bacteria have been found in such places as hot water tanks, showerheads, and air conditioning machinery. The disease is treated with antibiotics.

LENS IMPLANTS   See CATARACT

LEPTON   See ELEMENTARY PARTICLES

LEUKAEMIA *See* CANCER

LEVODOPA *See* PARKINSON'S DISEASE

LIFT

An upwards-acting force produced when the wing of an aircraft moves through the air.
See also AERODYNAMICS.

LIGASE *See* GENETIC ENGINEERING

LIGHT AND RELATIVITY *See* RELATIVITY

LIGHT-EMITTING DIODE *See* LCD AND LED

LIGHT MICROSCOPE *See* MICROSCOPE

LIGHT PEN *See* COMPUTER GRAPHICS

LIGHT POLLUTION *See* NOISE

LIGHT-YEAR

A measure of distance, even though it utilizes a unit of time in its name. It is the distance that light travels in one year, moving at a rate of just under 300,000 kilometres per second (186,000 miles per second). The light-year is a useful unit because it lets us avoid the kind of staggering number you are about to read.

In round numbers a light-year is a distance of 9.5 billion kilometres, or about 6 billion miles. Proxima Centauri, the star nearest our solar system, is 4.3 light-years away. For comparison, the moon is less than 1.5 light-seconds away from the earth.

LIGNASE *See* ENZYME

LIGNIN *See* ENZYME

LIGNITE *See* ENZYME

LIMESTONE *See* ROCK CLASSIFICATION

LINEAR ACCELERATOR *See* PARTICLE ACCELERATOR

LIQUID *See* MATTER

LIQUID CRYSTAL DISPLAY See LCD AND LED

LITHOSPHERE See PLATE TECTONICS

LOCAL GROUP

A cluster of galaxies. In addition to the Milky Way, our local group includes the Andromeda GALAXY, the Magellanic Clouds, and about two dozen other 'nearby' galaxies.

LONGITUDE See CELESTIAL COORDINATES

LOUDNESS CONTROL

A good high-fidelity set has a volume control and a loudness control. Both of them control the volume of the sounds coming out of the speakers, but the loudness control is more subtle. It takes into account the imperfections of the human ear and brain.

Assume that a recording of a string quartet (two violins, a viola, and a cello) is being played at about the same volume as a real quartet performing in the room. If you turn the volume control halfway down, it cuts the electric power to the speakers by about half. All the instruments should sound half as loud — but they don't. The balance is changed because your hearing apparatus doesn't hear low-volume sounds at each end of the scale (deep cello and high-pitched violin) as well as it hears low-volume middle-range notes. Turn the volume control still further down: the deep notes and the high notes will fade out completely, while the middle-range notes can still be faintly heard. (Imagine the mayhem done to 130 instruments playing Beethoven's Fifth!)

Now restore the volume control to normal and use the *loudness* control. Again, the electric power in the speakers is reduced. This time, however, the reduction is not at a straight-line rate but in a changing ratio called the Fletcher-Munson contour, after the two scientists who discovered and measured the uneven sensitivity of the human hearing system.

LOUDSPEAKER

A device that converts electrical impulses into sound vibrations. Most loudspeakers contain a permanent magnet surrounded by a movable electromagnet (*see* ELECTROMAGNETS). *The movable magnet,* called a *moving coil,* or *voice coil,* is attached to a cone of stiff paper. Each incoming electrical impulse magnetizes the coil, then demagnetizes it, on and off. The 'on' state sets up an attraction between the permanent magnet and the coil. With the 'off' the attraction ends. The paper cone, moved by the coil, makes one vibration of sound. In this way, a string of electrical impulses is converted into a string of sound waves. The earphones used with some radios are really tiny loudspeakers that use discs instead of paper cones.

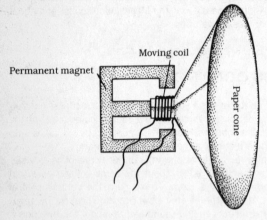

Strong electrical impulses cause big vibrations of the cone, producing loud sounds; weak impulses cause small vibrations, producing soft sounds. Many impulses per second produce high-pitched (soprano-like) sounds; few impulses produce low-pitched (deep) sounds.

A single loudspeaker cannot accurately reproduce both high- and low-pitched sounds; therefore, a high-quality speaker is really made up of two or more speakers. There is a *woofer,* with a cone 20.5 centimetres (8 inches) or more in diameter, to produce the deep booms and grunts of ominous timpani and basses. A *tweeter,* about 5 centimetres (2 inches) in diameter, specializes in the high notes issuing from violins, piccolos, and operatic heroines threatened by villains. Often there is also a *midrange speaker,* about 12.5 centimetres (5 inches) in diameter, to reproduce the middle-range tones faithfully.

A *crossover network* – a set of electrical circuits – apportions the proper electric frequencies to the appropriate speakers.

LOVASTATIN *See* CHOLESTEROL

LOW-DENSITY LIPOPROTEIN (LDL) CHOLESTEROL *See* CHOLESTEROL

LOW-LEVEL LANGUAGE See COMPUTER LANGUAGES

LUMEN (OF ARTERIES) See ARTERIOSCLEROSIS

LUNAR ECLIPSE See ECLIPSE

LYMPHATIC SYSTEM See CANCER

LYMPHOCYTES See IMMUNE SYSTEM

LYMPHOMA See CANCER

M

MACHINE LANGUAGE See COMPUTER LANGUAGES

MACROPHAGE See IMMUNE SYSTEM

MAGELLANIC CLOUDS

A pair of irregular galaxies, visible from the Southern Hemisphere; at a distance of 200,000 light-years, they are the closest galaxies to our GALAXY, the Milky Way.

'MAGIC BULLET' See CANCER

MAGMA See PLATE TECTONICS; ROCK CLASSIFICATION

MAGNET See ELECTROMAGNETS

MAGNETIC BOTTLE See NUCLEAR ENERGY

MAGNETIC FIELD See FIELD THEORY

MAGNETIC FIELD OF THE EARTH
See VAN ALLEN BELTS

MAGNETIC INK CHARACTER SORTER

A machine that reads letters and numbers printed with magnetic ink (*magnetic ink character recognition*) and then sorts the documents on which they appear. The most common use of the machine is to sort bank cheques. Each odd-looking character is designed on a 7-by-10 grid. A magnetic ink scanner in the sorter determines which of the 70 spaces is empty and which is filled with the magnetic ink. This is accomplished while the cheques race through the scanner without stopping, a dozen per second. Shown here is the magnetic ink code used for most sorting purposes. The symbols in the bottom line are codes for the beginning and end of such numbers as the customer's account number, the number of the bank, and the amount for which the cheque was written.

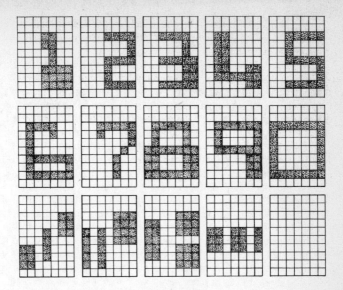

MAGNETIC RESONANCE IMAGING (MRI)

A noninvasive and painless system for examining and depicting the interior of the body. First, let us briefly review some invasive systems.

Surgery A totally invasive procedure, since it involves cutting open the body, probing, and closing up, with all the attendant risks.

X-ray examination (fluoroscopy) and photography Very much a 'handle with care' system, since X-rays in sufficient quantities are destructive to living tissue and cause cancer.

Ultrasound A safe process as far as is known; does not produce finely detailed images but is very effective in imaging gallstones and kidney stones and in following motion, as with the heart or with a fetus in the uterus.

CAT or CT scans (CAT and CT stand for *computerized axial tomography*.) Excellent images but uses weak X-rays, which can still add up.

PET scan (PET stands for *position emission tomography*.) Has the virtues of CAT and doesn't use X-rays; is also used to analyse chemical activity in an organ (especially the brain) without harming it.

And now for MRI, the technique using a *magnetic field* (see FIELD THEORY) to cause RESONANCE within atoms, producing an *image* by means of that resonance.

When an airline attendant invites you through a security checkpoint, you walk through a space surrounded by a wire loop, filled with an invisible electromagnetic field. This field has a certain frequency (a certain number of vibrations per second) in the same range as the frequency of metal atoms. The electromagnetic field causes certain metal atoms (e.g., keys or a concealed gun) to vibrate (resonate), and this resonance is picked up by a detecting instrument.

That whole apparatus is tuned to produce resonance in a broad field of substances — metals — but it could be adjusted to a specific substance such as gold, if we were on the lookout for stolen treasure.

The MRI technician invites patients to lie on a bed placed inside a powerful electromagnet, whose field is of the same frequency as hydrogen. This element is the most common in living tissues (water is hydrogen oxide, H_2O). Hydrogen is everywhere in the body, and it is detected everywhere, but its concentration varies greatly — very little in bone, more in muscle, more in certain glands and less in others, a great deal in blood, even more in urine, and so on. This varying concentration is detected, stored by a computer and analysed, and made into a computer-graphic picture. By changing the direction of the magnetic field continually, obtaining numerous pictures, a three-dimensional image is obtained. As far as is known, there are no harmful effects of magnetic fields; if that is so, MRI is totally noninvasive, and a boon indeed.

See also COMPUTERIZED AXIAL TOMOGRAPHY; ULTRASONICS.

MAGNETIC-TAPE RECORDING   See SOUND RECORDING

MAGNETOHYDRODYNAMICS (MHD)

The study of the effects of electric and magnetic fields on moving electro-conducting fluids, also known as *hydromagnetics*. MHD holds enormous promise for the future of electric power generation. Consider first how most electric power is generated nowadays (*see* GENERATORS AND MOTORS, ELECTRIC):

A flame (oil, gas, or coal) or a nuclear reactor → heats water → into steam → which turns a turbine → which turns a generator → which generates an electric current.

Five arrows, five transfers of energy, five bribes! No machine is 100% efficient, so a little bribe must be paid, a little energy dissipated uselessly, during each transfer. Overall efficiency is about

40% (see ENTROPY). Now look at the transfer sequence in a magnetohydrodynamic generator:

A flame → heats a liquid or gas that can conduct electricity → which generates electric current. Overall efficiency is about 60%.

Where are these ideal machines? All that stands in their way is technical refinement (which reminds us of the glum gentleman who remarked. 'All that stands between me and happiness is misery'). Look at the generator diagram on page 134. Its essence is this: when a *magnetic field* (around the magnet) moves across a stationary conductor of electricity (the coil of wire), it sets up an electric current. The same effect can be achieved by *moving the conductor* across a stationary magnetic field. The conductor needn't be a coil of wire, either. It can be a hot fluid (a gas or liquid) that conducts electricity. The fluid flows through a pipe surrounded by a strong magnet. The current is picked off by collectors (electrodes) at the two ends of the pipe.

What refinements are needed?

1. The efficiency of heat transfer to the fluid must be improved; too much is lost at present.
2. The fluid has to be heated to 4000°C (over 7200°F) to be sufficiently conductive (ionized).
3. The pickup needs to be more efficient; at present, too much is lost.
4. And there are other difficulties.

But MHD research is continuing.

MAGNETOSPHERE   See VAN ALLEN BELTS

MAGNETRON   See MICROWAVE OVEN

MAGNITUDE

The brightness of a star, planet, or other astronomical object, measured on a scale of magnitudes. For complicated historical reasons the scale assigns increasingly negative numbers to the brighter objects and increasingly positive numbers to the dimmer ones. Thus the sun, the brightest object in our sky, has a magnitude of −26; the full moon is about −12; Sirius, the brightest star we see, is −1.5; Polaris, the Pole Star or North Star, is +2; and dim stars visible only in the largest telescopes have magnitudes of +20 or more.

A difference of one magnitude represents an actual brightness difference of about 2½ times (2.512, to be exact). That makes a first-magnitude star about 2½ times as bright as one of second magnitude and over six times as bright as a third-magnitude star (2½ times 2½). The differences in magnitude add up (or, rather, multiply) very quickly: a difference of only five magnitudes represents a difference of 100 times in brightness.

So far, we've touched an *apparent magnitude* — how bright a star looks to us here on the earth. But that depends on how brightly the star *really* shines and on how far away it is (*see* INVERSE SQUARE LAW). Roughly put, a nearby dim star may *look* brighter than a very bright star very far away. Astronomers and others who must know the brightness of stars precisely perform a calculation that, in effect, moves *all* stars to an imagined equal distance from the earth. The magnitude that a star would have at that distance is called its *absolute magnitude;* this allows direct comparison of the real brightness of one star with the real brightness of any other star.

MAINFRAME   See COMPUTER

MALIGNANT TUMOUR   See CANCER

MANUFACTURING PROCESSES

Various methods of mass-producing objects out of raw materials. These methods have impressive technical names (later) but can be described as cutters, tubes, moulds, and rollers.

Cutters Biscuit cutters stamp an outline into a sheet of dough. A fancier kind stamps not only an outline but a design on the dough.

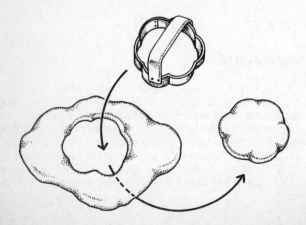

MANUFACTURING PROCESSES

The cutters at the Royal Mint are of two kinds: one kind stamps discs out of sheet metal; the other punches designs on the discs. The result is pennies, 5p pieces, 10p pieces, and so on. The forms that produce these shapes and designs are called *dies*. The machines that do the stamping are called *die presses, punch presses,* or *hydraulic presses* (if they work by hydraulic pressure).

Try a variation of this idea. Suppose the die is made with a bent shape — a curve or angle. A flat sheet, if struck by such a die, would be bent into that shape. That's how, in a car factory, wings and tops are made. First, flat sheets of steel are punched out by presses that form a flat outline of the piece. Then other presses bend these outlined pieces into three-dimensional forms.

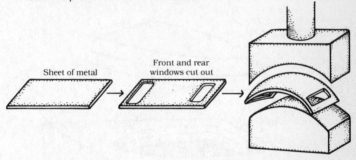

Tubes The shape of a tube's nozzle determines the shape of the substance, say toothpaste, that comes out of the tube. This principle, called *extrusion*, applies to the manufacture of many kinds of long shapes: rods, strips, tubes, railings, and so on. Molten plastic or metal is forced by a plunger, called a ram, through a shaped hole in a hard steel die.

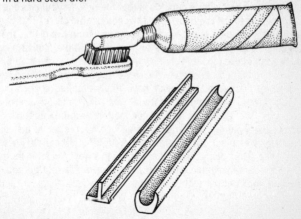

Variation: suppose that instead of pushing the raw material through a die, you pull it. By pulling a pointed rod through a die hole slightly smaller than the thickness of the rod, then through a still smaller hole, and again, and again, the thickness is gradually reduced until the rod has been drawn out and become a wire. This is the *wire-drawing* process.

Moulds Pour liquid jelly into a shaped container, a mould. The jelly cools and hardens into a solid object bearing the shape of the mould. This is the principle of *casting* (Old Norse *kasten,* to throw or pour). An enormous number of manufactured objects are castings (e.g., iron manhole covers, electric irons, chocolate bars, wedding cakes).

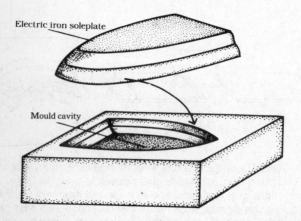

A problem with simple moulds is that the process imposes one flat surface (the top of the pour) onto the shape of the piece. This can be avoided, when necessary, by making the mould in two pieces that fit together, each with half of the form. A drilled hole (called a gate) allows the molten material to flow into the interior of the mould. Let it cool and harden, take the two halves of the mould apart, and *voilà!*

Using this method you can't have finely detailed shapes in the mould, because the molten material won't flow into tiny nooks and crannies. The solution is to *force* the molten material into the mould by putting a plunger or compressed air behind it. That is, instead of pouring, we *inject* the molten material. This is called *injection moulding*. Most plastic objects are made this way. A related method, used by dental technicians, is called *centrifugal casting* and will provide your dentist with a welcome new lecture topic, in lieu of politics.

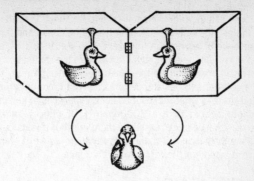

Rollers Forcing a soft material such as pasta dough against a pair of parallel rollers turning forwards in opposite directions will result in a flat layer of dough the thickness of the distance between the rollers. In a steel mill, molten-steel is first poured into moulds, forming square chunks of steel called *ingots*. The red-hot ingots are then passed through rollers that squeeze them, thinner and thinner, until the desired thickness of *sheet steel* is obtained.

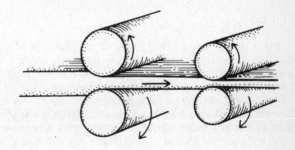

MARBLE See ROCK CLASSIFICATION

MARITIME AIR MASSES, POLAR AND TROPICAL See AIR MASS ANALYSIS

MARSH GAS See ENERGY RESOURCES

MASKING AGENTS (DRUGS) See STEROIDS

MASS

A measure of the amount of MATTER in an object. Mass (Greek *maza*, lump) is commonly determined by weighing machines, such as bathroom scales and chemical balances. These instruments measure the gravitational attraction between the object and the earth. In the absence of gravitation (for example, adrift in space), all objects have zero *weight,* even though their *mass* is the same as on earth.

However, mass can be measured by another property of matter, *inertia,* which is its resistance to a change in motion or nonmotion. Inertial mass is what makes it easier on your toes to kick a 50-gram (5-ounce) pebble than a 3000-gram (105-ounce) brick. For the same reason (the more mass, the more inertia), it is harder to stop a 20-tonne lorry than a 2-tonne car, anywhere on earth — and equally so on the moon, on Jupiter, or in empty space.

MASS, RELATIVE   See RELATIVITY

MASS DEFECT   See NUCLEAR ENERGY

MASS EQUIVALENCE   See NUCLEAR ENERGY; RELATIVITY

MASS STORAGE   See COMPUTER

MATTER

'Is this a dagger which I see before me?' asks Macbeth. Well, it is or it isn't, depending on whether it passes the simple test for matter: does it occupy space? To which we can add as a further check: does it have MASS and inertia? A true dagger, furthermore, is a *solid,* with a given shape and volume. The molecules in solids maintain a constant position in relation to one another, which is why you can count on finding a metre rule's 50-centimetre mark right where it is, always between the 49 and 51, not wandering from time to time. Most solids (for example, rocks and metals) are crystals, with molecules arranged in orderly, geometric form.

Solids turn to *liquids* at certain temperatures. In the liquid state, their molecules are free to wander, enabling a drop of blue ink to diffuse throughout a bucketful of clear water and allowing the Gulf Stream, heated in the sunny Caribbean, to eventually warm the coast of Labrador. There is, nevertheless, enough attractive force between the molecules of liquids to keep them in contact with each other, which is why a cupful of coffee continues to be a cupful while you stir it. But the attractive force is disrupted by sufficient heating. The molecules become increasingly agitated in their motion. Finally they move far enough apart to break free and leap out of the liquid, in the process called *evaporation,* forming a *gas*. When molecules in the gaseous state collide, they have sufficiently violent agitation to bounce off each other, rather than succumbing to the intermolecular attractive force.

MECHANICAL ENERGY *See* ENERGY

MEDIA (OF ARTERIES) *See* ARTERIOSCLEROSIS

MELTDOWN *See* NUCLEAR REACTOR

MEMORY, COMPUTER *See* COMPUTER

MERIDIAN *See* CELESTIAL COORDINATES

MESON

A subatomic particle consisting of a quark and an antiquark.
See also ELEMENTARY PARTICLES.

MESSENGER RNA *See* DNA AND RNA

METAL CASTING *See* MANUFACTURING PROCESSES

METAL DETECTOR *See* MAGNETIC RESONANCE IMAGING

METAL ROLLING *See* MANUFACTURING PROCESSES

METAMORPHIC ROCK *See* ROCK CLASSIFICATION

METASTASIS *See* CANCER

METEOR *See* SOLAR SYSTEM

METEORITE *See* SOLAR SYSTEM

METEOROID *See* SOLAR SYSTEM

METEOROLOGY *See* AIR MASS ANALYSIS; PHYSICS

METHANE *See* ENERGY RESOURCES

MeV *See* ELECTRON VOLT

MHD *See* MAGNETOHYDRODYNAMICS

MICROBES *See* ANTIBIOTICS

MICROBIOLOGY See MOLECULAR BIOLOGY

MICROPROCESSOR See COMPUTER

MICROSCOPE

An instrument for producing enlarged images of small objects. Microscopes range from simple single-lens magnifiers to complex electronic instruments costing hundreds of thousands of pounds.

Microscopes work like showerheads. A small circular input — water flowing through a pipe — is spread into the shape of an expanding cone. Get in the way near the small end of the cone, and you're hit with a small disc of water. Farther away from the showerhead, you're hit with a larger disc. Notice the same effect with a film or slide projector. The greater the distance to the screen, the larger the image. A projector is really a projection microscope.

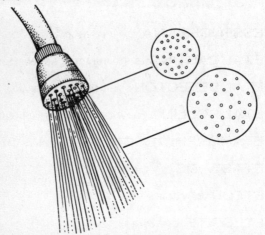

Look at a small drop of water lying on a leaf, a patterned work surface, or some other nonabsorbing surface. The transparent, *convex* (bulging) water drop causes light to bend and spread out in cone-shaped form, thus magnifying the details of the underlying surface. A bead or some other convex piece of glass does the same job, with the additional virtue of being permanent.

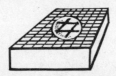

An ordinary magnifying glass is a good example of a *simple microscope* (it has only one lens). It can magnify clearly perhaps 10 to 20 times, but is limited to that. The *compound microscope* is a great improvement. It has sets of convex lenses at each end of a tube. The first set of lenses (the objective) forms an enlarged image of the object, and the second set (the eyepiece) enlarges that image.

Both simple and compound microscopes make use of light waves and are therefore called *light,* or *optical, microscopes.* The best of these are limited to a magnifying power of about 2000 times (if pushed higher, they get fuzzy) because they can't form images of objects that are smaller than the light waves. It's like trying to draw a picture of a spider's web with a thick crayon.

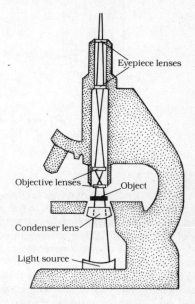

To get higher magnification we must use something even smaller than light waves — beams of electrons. Electron beams can't be used in optical microscopes because, unlike light waves, they are not bent by glass lenses. But electromagnetic fields can bend them, to form images. In *electron microscopes,* ring-shaped electromagnets act as lenses with beams of electrons, spreading them out into cones, like the cone of water in the showerhead. However, the images formed by electrons are invisible to human eyes. To make them visible, the images are formed on a glass screen, coated like a television tube with material that glows when struck by electrons.

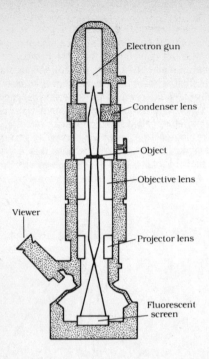

There are two basic kinds of electron microscopes, both capable of magnification of 1 million times or more. In *transmission electron microscopes,* the electron beams are transmitted through extremely thin slices of the material being examined. In *scanning electron microscopes,* a thin beam of electrons sweeps back and forth over the specimen. The electrons scan the material without penetrating it, so there is no need to slice it thin. This makes it possible to examine and photograph very small living objects. The images produced by this instrument have a strongly three-dimensional character.

See also RESOLVING POWER.

MICROSURGERY

A relatively new development in surgery, in which operations are performed under specially designed microscopes. The most publicized use of microsurgery is in reattaching fingers, toes, and even whole limbs lost in accidents, but less dramatic microsurgical operations are done routinely. In the brain, for example, tiny weakened arteries are repaired and clogged arteries are cleared to prevent stroke.

Aside from the operator's skill, the technique is made possible by the use of miniaturized instruments. Observing through the microscope and using finger movements that are barely perceptible, the microsurgeon repairs, cuts, transplants, and reconnects tiny muscles, along with nerves and blood vessels that may be no thicker than a hair.

MICROWAVE OVEN

An oven that heats food by means of very short radio waves called *microwaves*. The waves are generated within the oven by an electron tube called a *magnetron*. They cook food much faster than ordinary ovens.

In an ordinary gas or electric oven, heat from a flame or from a heated coil strikes the cooking vessel, which becomes hot. The heat is conducted into the outside of the food and from there to the inside, progressively. Microwaves, however, pass through cooking vessels made of pottery, china, plastic, and even paper without heating them, the way ordinary radio waves pass through most substances. But the microwaves cause water molecules to vibrate rapidly, and their friction produces heat. Since all foods contain some water throughout, a microwave oven cooks food quickly inside and out.

See also RESONANCE.

MICROWAVES   *See* MICROWAVE OVEN

MIDRANGE SPEAKER   *See* LOUDSPEAKER

MILKY WAY

The GALAXY in which our solar system is located. The light of thousands of millions of stars that make up the Milky Way cause it to appear as a pale band of light in the night sky.

MINERALOCORTICOIDS   *See* STERIODS

MINI-SATELLITE   *See* DNA FINGERPRINTING

MINOR PLANET   *See* SOLAR SYSTEM

MODEL, SCIENTIFIC   *See* SCIENTIFIC TERMS

MODEM *See* INTERFACE

MODERATOR *See* NUCLEAR REACTOR

MOLECULAR BIOLOGY

The study of life at its fundamental level — the complex molecules on which it is based. The field includes the *nucleic acids, proteins,* and *genetic engineering*. This young science draws from advances in research in *biochemistry* and *biophysics* and overlaps both to some extent.

The history of this major field of science is, in an intriguing way, shaped like another field, atomic theory. In both fields a basic question — What is the world made of? What is life? — gave rise at first to several seemingly simple answers, which on further examination multiplied into numerous bewilderingly complex explanations. These, on still further examination, coalesced into a comparatively few basic, apparently fundamental concepts about what, when, how, and where, with some tentative pokes (by theologians and philosophers) into who and why.

Thus the early classification of matter into four 'elements' - earth, air, fire, and water — proliferated into listings of thousands of chemical compounds and reactions, until a unifying group of concepts, the atomic theory, began to show through. Today, scientists seem to be enticingly close to *the* fundamental unification point as suggested by the GRAND UNIFIED THEORY: the material universe seems to consist of quarks, leptons, and their antiparticles, joined by two forces (perhaps two aspects of one force).

Back to the facts of life. Molecular biology is the area of science that studies living things in their most mechanistic aspect: assemblages of molecules, composed of chemical elements and compounds, interacting with one another and with their environment. Sounds rather soulless, as indeed it is, but it does get down to one kind of fundamental: the pinpointed scientific study of life.

Molecular biology embraces several subdivisions and overlaps others. Let's look at some of the more prominent ones.

Organic chemistry The study of carbon compounds. The basic substances in all living things are proteins, and the basic elements in all proteins are carbon, hydrogen, oxygen, and nitrogen.

Molecular genetics The study of genes, the carriers of hereditary information in the cells. Genes are mainly long strings of molecules called nucleic acids. Their arrangements and interactions add up to the *physical* phenomenon we call *life*.

Pharmacology The study of the effects produced by chemicals, especially those classed as drugs, on living things.

Toxicology The study of poisons (toxins): their effects, detection, isolation, and identification. This field could be considered a subdivision of pharmacology.

Biochemistry Closely related to and overlapping molecular biology. Such distinctions as do exist are not significant to the nonspecialist.

Microbiology The study of bacteria, viruses, and other microorganisms, especially their life cycles and effects at the molecular level.

Physiology The study of life processes, such as circulation, digestion, and metabolism. Here, too, the focus becomes increasingly minute, down to the molecular level.

Pathology The study of the causes and effects of disease, again with greater and greater emphasis on the molecular level.

See also DNA AND RNA; GENETIC ENGINEERING; PROTEINS AND LIFE.

MOLECULAR GENETICS *See* MOLECULAR BIOLOGY

MOLECULAR THEORY *See* SCIENTIFIC TERMS

MOMENTUM, MEANING OF *See* UNCERTAINTY PRINCIPLE

MONGOLISM *See* DOWN'S SYNDROME

MONOMER *See* CHEMISTRY

MONONUCLEOSIS *See* HERPESVIRUS DISEASES

MONOPHONIC SOUND *See* QUADRAPHONICS

MONOUNSATURATED FATS *See* CHOLESTEROL

MOSAIC DISEASES *See* VIRUSES

MOTION, LAWS OF *See* NEWTON'S LAWS OF MOTIONS

MOTION, RELATIVE *See* RELATIVITY

MOTOR NEURON DISEASE
See NEUROMUSCULAR DISORDERS

MOTORS, ELECTRIC
See GENERATORS AND MOTORS, ELECTRIC

MOVING COIL See LOUDSPEAKER

MRI See MAGNETIC RESONANCE IMAGING

MULTIPLE SCLEROSIS (MS)
See NEUROMUSCULAR DISORDERS

MULTIPLE STARS See BINARY STAR

MUON See ELEMENTARY PARTICLES

MUSCULAR DISORDERS
See NEUROMUSCULAR DISORDERS

MUSCULAR DYSTROPHY See NEUROMUSCULAR DISORDERS

MUSICAL ACOUSTICS See ACOUSTICS

MUTATION See DNA AND RNA

MYASTHENIA GRAVIS See NEUROMUSCULAR DISORDERS

MYELIN SHEATH See NEUROMUSCULAR DISORDERS

MYOCARDIAL INFARCT See HEART ATTACK

MYOCARDIUM See HEART ATTACK

NALOXONE See ALZHEIMER'S DISEASE

NATURAL SELECTION See EVOLUTION

NEBULA

The term (from Latin *nebulosus,* cloud) originally referred to fuzzy-looking astronomical objects that early telescopes did not focus as individual stars. The so-called spiral nebulas were among these objects. Later it was determined that these nebulas lay beyond the Milky Way and that they were, in fact, separate galaxies. Today the term is applied mainly to zones of dust and gas within the Milky Way, but old usages die slowly. For example, the Andromeda galaxy is often called the Andromeda nebula.

NEOPLASM See CANCER

NERVE-MUSCLE DISORDERS
See NEUROMUSCULAR DISORDERS

NEUROMUSCULAR DISORDERS

A number of disorders in which body muscles, the nerves that control them, or both, malfunction. Among the better-known of these are the following, all of unknown cause:

Myasthenia gravis A disorder in which muscles become weak and easily fatigued. It appears mainly between the ages of 20 and 40 and twice as often in women as in men. Muscles in the head are often affected, so there may be drooping of eyelids, double vision, and difficulty in swallowing; a special danger arises if the muscles connected with breathing become involved.

The fault appears to be in the *neuromuscular junctions,* the places where 'orders' to contract pass from nerves to muscles. The 'order' is a chemical reaction, and in myasthenia gravis there is an imbalance in the amounts of the reacting chemicals. Drugs are available that help to restore the balance and thus improve the working of the muscles. (Ironically, these drugs, called *cholinesterase inhibitors,* are also used in chemical warfare as nerve gases.)

Motor neuron disease A group of diseases involving degeneration of nerves that control muscular contractions, occurring mainly in people over 40. The muscles weaken and waste away. Physiotherapy helps in keeping the muscles working, but the condition is usually fatal within a few years. *Amyotrophic lateral sclerosis* is the commonest form.

Multiple sclerosis A disease of the nervous system (also called *MS* or *disseminated sclerosis*) affecting mainly young adults. Early symptoms may include lack of feeling in some parts of the body, weakness and clumsiness, problems of vision, and apathy. The disease progresses very slowly – the average duration is 25 years or more, though there is great variability – and there are usually periods of remission, when the symptoms lessen or disappear. In time, however, the person may be permanently disabled.

Nerves affected by the disease lose some of their outer insulation-like sheath, a fatty material called *myelin;* the nerves harden (*sclerosis* is from the Latin for 'hardening'). The evidence indicates that MS is an *autoimmune* disease, that is, one in which the body is attacked by its IMMUNE SYSTEM. In the case of MS, it is the myelin-producing cells that are attacked.

Physiotherapy and psychological support are important in helping the patient. Recently treatment with a synthetic protein, *copolymer 1 (Cop 1)*, has raised hopes for controlling the effects of MS.

Muscular dystrophy A group of inherited disorders, with a progressive weakening and wasting of muscles. It is believed that the fault lies in the muscles themselves, rather than in the nerves. Physiotherapy helps to improve the working of the affected muscles.

One type, affecting only males, begins before the age of 3 and ends in death before the age of 30. Another type, beginning in adolescence, affects both sexes and may range from relatively mild to disabling.

NEUROTRANSMITTER   See PARKINSON'S DISEASE

NEUTRINO   See ELEMENTARY PARTICLES

NEUTRON

A subatomic particle that is electrically neutral; that is, it has no charge. Neutrons, along with protons, make up the nuclei of atoms, with one exception – the nucleus of the simplest form of the lightest element, hydrogen, has one proton and no neutrons. Protons and neutrons are called *nucleons* because they are the main components

of atomic nuclei.
See also ELEMENTARY PARTICLES; NUCLEAR ENERGY.

NEUTRON STAR *See* STELLAR EVOLUTION

NEWTON'S LAWS OF MOTION

What makes an arrow keep going after it leaves the bowstring? In the 4th century BC Aristotle suggested this: the arrowhead presses the air in front of it; the pressed air flows around back to the tail and pushes it forward. Before you raise your eyebrows, reflect kindly on Aristotle's paucity of material experience. He had never seen, even on TV, a spaceship circling the earth in airless space, with nothing squeezing from behind and a prospect of months, even years, of engineless flight ahead. What indeed keeps a spaceship going after its fuel is spent? Or even a tennis ball after it leaves the racket? Or any projectile that isn't simple dropped, plop, into the waiting arms of gravity?

Science didn't have to wait for the vacuum conditions of space flight to disprove Aristotle's guess about the flight of arrows. Isaac Newton (1642–1727), English scientist and mathematician extraordinaire, creator of CALCULUS, deviser of the laws of universal GRAVITATION, investigator in the fields of optics, theology, mathematics, astronomy, chemistry, mechanics, dynamics, and – alas – the occult, did it without vacuums. In propounding his three laws of motion, he put into concise and mathematical form the basic ideas of what makes things move, what changes their motion, and why motions always begin in pairs. These are the laws:

1. *A body at rest remains at rest, and a body in motion remains in motion in a straight line unless acted upon by a force.*

    This book would remain stationary until you chose to do something about it. Arrows would continue to fly forwards forever in a straight line were it not for air friction, which slows them, and gravity, which brings them down to earth. Planets, travelling through empty space, continue to move forwards, but *not* in a straight line, because the constant planet-sun gravitational force acts as a tether that bends the straight line into an almost-circle, an ellipse.

2. *The effect produced on a body by a force depends (a) directly on the amount of that force and (b) inversely on the mass of the body.*

    To put it more plainly: (a) kick a football twice as hard and it will go twice as far, and (b) kick a double-weight football (don't quibble) and it will go only half as far.

3. *For every action there is an equal and opposite reaction.*

To climb a ladder, you must push down on the rungs. To walk north, your feet must push south (and if as on ice or a highly polished floor they can't push south, you don't get to walk north). For an aeroplane to fly east, it must push air west. To manoeuvre in space, with no air to push against, a spacecraft has to carry its own 'push against' material: tanks of compressed gas, which is let out in little puffs: a puff out of the left rear nozzle nudges the tail to the right; a rearward puff increases the forward speed; a forward puff decreases the forward speed (puts on the brakes); and so on.

Newton's three laws seem quite simple and commonsensical — which they are. Yet when he organized them into algebraic equations, they served as the basis for calculating the entire gravitational force between the earth and moon. From that calculation he worked out the law of universal gravitation, which applies to any two masses in the universe.

NICAD BATTERY   See BATTERY (CELL)

NITROGEN MUSTARDS   See CANCER

NITROGLYCERIN   See ANGINA PECTORIS

NMR   See ANALYSIS

NOISE

Noise is more than meets the ear. Common noise refers to undesired, incoherent sound — the rumble of traffic, the buzz of a cocktail party — that interferes with the comprehension of desired sound, such as speech or music. By extension, noise has come to mean any interference of a random, unspecific nature. A poorly constructed refrigerator motor emits *radio noise* waves that pepper your radio programmes with crackles or scatter 'snow' on your television screen.

Astronomers are having serious problems with *illumination noise* as growing cities expand towards once-isolated observatories. The light scattered by street lamps and electrical signs (sometimes called *light pollution*) throws a glow on the night sky, overwhelming the light of dim stars, galaxies, and nebulas.

The strength of a desired signal, such as a radio programme, compared to the strength of an undesired signal, such as radio static, is called the *signal-to-noise (S/N)* ratio. Well-made radio receivers can operate with a low S/N ratio — that is, they accept coherent radio waves but reject radio noise of almost the same strength. Inferior receivers operate on a high S/N ratio — they can produce uncluttered sound only if the signal strength is much higher than the noise.

NOISE INSULATION ENGINEERING   See ACOUSTICS

NONRENEWABLE ENERGY SOURCE

A source of energy that is depleted as it is used. Coal is an example. *See also* ENERGY RESOURCES.

NORADRENALINE   See HORMONE

NOVA

A faint, usually unseen star, whose light may brighten by a factor of 10,000 or more in a few days and then fade slowly. A few novas (from Latin *novus,* new) appear each year. Apparently these outbursts occur when gases from the larger member of a BINARY STAR fall onto the smaller member, setting off a nuclear explosion, with its attendant brilliance. The larger star is not affected significantly because the lost material represents only a minute fraction of its mass.

A far more brilliant, and very rare, outburst of light, the *supernova,* is caused by the explosion of a star.

See also STELLAR EVOLUTION.

NUCLEAR ENERGY

The energy released by the nuclei of atoms when they undergo certain changes (to be described shortly). Formerly called atomic energy, the term *nuclear energy* is more precise, since only the nucleus of the atom is involved. Nuclear energy is the oldest *and* the newest form of energy: the oldest because in its natural form it was involved in the first moment of cosmological history, the newest because its artificial form was first produced in quantity in the 1940s by Enrico Fermi and others in the Manhattan Project.

Nuclear energy is generated within all the stars of the universe; it is the source of our sunlight and consequently is the primary energy basis for all life on earth. Misused, its artificial form can bring about the extinction of all life on earth.

The lightest, smallest, simplest nucleus is that of ordinary hydrogen, which consists of a single proton and nothing else. The heaviest, largest, most complex nucleus of a natural element, uranium 238, consists of 92 protons and 146 neutrons ($92 + 146 = 238$).

NUCLEAR ENERGY

Protons repel each other with tremendous force. Perhaps you have had the experience of trying to press together two strong magnets, with like poles opposed, repelling each other. In an equal mass of protons the repulsive force is millions of times greater.

Then what holds the nucleus together? *Binding energy,* a holding operation involving the neutrons and evanescent particles called *gluons.* And just as the repulsive force within the nucleus is enormous, so is the binding energy that keeps the nucleus from flying apart. Every nucleus of every atom (except simple hydrogen) would fly apart but for the binding energy within the nucleus. It is this binding energy that, when released slowly and under control, produces heat that powers steam-driven electric generators in nuclear power plants. It is this binding energy that, when released all at once, delivers the smashing destructiveness in a nuclear bomb.

Does this mean that we could obtain nuclear energy from almost any substance? Theoretically yes; practically no — because the process of triggering the energy release would in most cases require more energy than the energy released. But nuclear energy is being obtained right now, in two ways:

1. *Fission* — encouraging the nuclei of very heavy atoms (uranium and plutonium, usually) to break apart into approximately equal parts (barium and krypton, usually)
2. *Fusion* — forcing very light nuclei (of hydrogen) to combine, forming somewhat heavier nuclei (of helium)

Before we look at these processes, some provisos and ahems:

Complexity The nucleus is much more complex than the model of simple magnets repelling each other.

Isotopes The nuclei of any specific element have the same number of *protons,* but the number of *neutrons* is variable, within a range. The various forms are called *isotopes.* The common isotope of hydrogen has one proton. Its isotope *deuterium* also has one proton, plus one neutron. Its isotope *tritium* (TRIT-e-um) has one proton and two neutrons. As we saw, the common isotope of uranium, U-238, contains 92 protons and 146 neutrons (92 + 146 = 238). Another

NUCLEAR ENERGY

uranium isotope, U-235, contains 92 protons and 143 neutrons (92 + 143 = 235).

Energy is real Although it has no shape, size, or colour, energy is real, because it has *mass equivalence*. Meaning what? A nucleus in the most common isotope of helium consists of two protons and two neutrons. Separately, these four particles weigh a total of 4.0320 units. Together, combined into one nucleus, they weigh 4.0016 units! The loss of weight (mass) is called *mass defect* and is due to matter that turned into energy — the binding energy that keeps the nucleus from flying apart. Binding energy is derived from the matter in separate protons and neutrons. (And energy can be reconverted into matter, but that's another story.) *Matter and energy are interchangeable,* or *equivalent*.

Albert Einstein in 1905 (at the age of 26!) worked out an equation that describes the equivalence of matter and energy: $E = mc^2$. Stated in words; the energy (E) obtained from the conversion of a certain amount of matter (into energy) is equal to the mass defect (m) in the conversion multiplied by the square of the speed of light (c) in centimetres per second (almost 30,000 million). In simple terms, a paper aeroplane ticket, if completely converted into energy, could propel an airliner several thousand times around the world.

You might suppose, then, that nuclear energy — conversion of matter into energy — is, or could be, an endless resource, since paper for aeroplane tickets is indeed plentiful. Restraining our exuberance, we briefly examine the present situation regarding fission and fusion, to find out what is holding back the wheels of progress.

Fission First, an analogy (but only analogy!) to help visualize fission. A medium-sized oil drop is floating on water. Add a tiny oil droplet; it is taken in and the oil drop becomes bigger, held together by the attractive force among the oil molecules. Add another oil droplet and still another — there's a certain larger size at which another phenomenon occurs: the next tiny oil droplet becomes (to pile analogy upon analogy) the straw that breaks the camel's back. The large drop breaks — it undergoes fission — into two medium-sized drops. What's more (you couldn't know this — it has to be measured), the two medium-sized drops require less total attractive force to retain their form than the one large drop did! And what happened to the leftover attractive force? It couldn't just disappear; it agitated the water molecules and shook up the water surface.

Now let's consider the real thing. Uranium is a heavy element and comes in ten isotopes. One of them, U-235 (92 protons, 143 neutrons) has an unstable nucleus, uneasily held together (the big drop). It is easily broken apart if hit by a neutron (tiny oil droplet).

NUCLEAR ENERGY

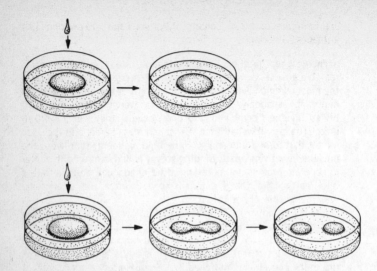

Then it breaks apart into two nearly equal-sized nuclei, krypton and barium, liberating a great deal of binding energy (resulting mainly in heat). Also left over, in this nuclear arithmetic, are several neutrons that are ejected during the turmoil. These released neutrons, if they strike other U-235 nuclei, perform the same fission-triggering function on them. They in turn trigger more fission, and so on. The whole sequence is called a *chain reaction* (though branching is perhaps more descriptive). If the U-235 fuel supply is arranged to produce a sequential, moderated release of energy, we have controlled fission, fine for delivering heat to the boilers of steam-driven turboelectric generators. If the fuel supply is arranged to undergo fission and deliver all its energy instantaneously (in about a millionth of a second), we have, alas, a nuclear fission bomb.

Let us ignore the bomb (if only we could) and ask what's the problem with controlled, useful, economical nuclear energy. Why do

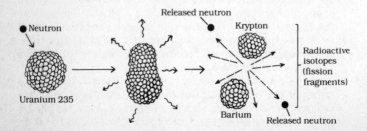

we continue to burn coal and oil in conventional engines and furnaces, polluting the air, rather than consuming aeroplane tickets (or their more practical equivalent, U-235) in nuclear reactors? Mainly for the following necessary but deplorable reasons:

Unquiet ashes Fission is not a simple split-up, like the oil drop; it actually involves a series of nuclear changes following the major heat-yielding split. At each step in the series, energy is given off in the form of radioactive particles and rays, some highly dangerous, until the 'ashes' have quietened down into safety. Some of these steps take thousands of years, during which time the *radioactive wastes (radwastes)* are lethal. Nobody has yet found a 100% surefire method of storing them, safe against corrosion and leakage of containers, safe against breakage during earthquakes, and so on. To those who say, 'But surely scientists will ... ' there is the chilling reply. 'But what if they don't?'

Another problem: a nuclear reactor becomes radioactive over its useful life of 30 years or so. It must eventually be disassembled, and its parts handled like other radwaste, or it must be entombed in concrete; either method is enormously expensive, costing perhaps one-third or more of what it cost to build the plant.

Unsafe safety valves The useful end product of nuclear fission is heat, to make the steam to run the turbine, and so on. The heat must flow unimpeded from its source, nuclear fuel, to its delivery point, the boiler where water is turned to steam. Any traffic jam means a backup of heat, a buildup of fuel temperature, which rises higher and higher, until it reaches *meltdown* temperature; then, as almost happened at the Three Mile Island reactor near Harrisburg, Pennsylvania USA, in 1979, and perhaps did happen in Chernobyl, in the Soviet Union, in 1986, the highly corrosive molten material may eat its way through the nuclear reactor, through its concrete shield, into the supporting bedrock (where it may flow into underground streams), and into the boilers where water is turned into radioactive steam, and be blasted far and wide into the atmosphere.

In 1988, nine years after the Three Mile Island reactor accident, engineers were still working on the $1000 million cleanup of the ruined reactor. They found that there had indeed been a massive meltdown of the fuel, which burnt through inner parts of the reactor. It collected in a solidified pool at the bottom but did not escape to the outside.

Yes, there are many safety devices in a nuclear power plant.

Yes, there have been numbers of near-disasters.

In fact, in mid 1988 the completed but unused Shoreham nuclear power plant on Long Island, New York, was abandoned because authorities were unable to devise a workable emergency evacuation plan for the people of Long Island.

Fusion Eureka! Here, at last, is a nuclear energy process with numerous virtues:

- Its principal raw material, the hydrogen isotope deuterium, is as close as the nearest body of water, H_2O. Only one hydrogen atom in 7000 is deuterium, yet complete fusion of the deuterium in an Olympic-size swimming pool could keep a city of 250,000 people in electricity for a year.
- The end product of fusion is helium, a stable, harmless gas used in balloons.
- The process of nuclear fusion has been tested and found successful since the beginning of time, as far as we know, in all the stars.

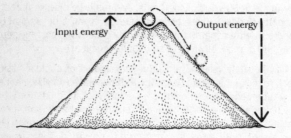

Hydrogen is fused by various steps into helium, with the release of enormous quantities of heat and radiant energy.

So what's stopping us from using it? Fusion involves some very large problems, and to understand them we must back up a bit. Fusion requires the forcing together of hydrogen isotopes to form helium nuclei. This requires energy to overcome nuclear repulsion, but after that happens far more energy is released than was needed to start it. Here's an analogy: a heavy ball rests in a cup-shaped hollow at the top of a mountain. If it rolls down the mountain, the ball can release a lot of stored (potential) energy. But first we must put it into a condition to roll, by expending a little energy to lift it to the rim. Likewise, before we can release the energy of nuclear fusion, we must put nuclei in a condition to fuse. This requires expending energy for two purposes:

1. To separate the nuclei from their surrounding electrons, so the nuclei can be brought close together. This stripped-apart

substance is a dense mixture of nuclei and free electrons: it's called *plasma* (not to be confused with blood plasma).
2. To slam the nuclei together so violently that they stick together (fuse).

Steps 1 and 2 can be achieved at a temperature of 100 million degrees Celsius. (Nuclear fusion is called a *thermonuclear* reaction.) This temperature exists all the time — in the interiors of the sun and other stars — and there are ways of producing it on earth, by electromagnetic and laser devices.

But first, a note to the sharp-eyed reader: yes, fusion, *assembling* nuclei, is the opposite of fission, *splitting* nuclei. Yet the fusion of light nuclei (e.g., hydrogen) and the fission of heavy nuclei (e.g., uranium and plutonium) are both energy-*releasing* processes. It's that matter of light and heavy and differences of binding energy, which would require a distressing number of words and equations to explain fairly.

Coming back to the main idea: where are the fusion reactors that will usher in the Golden Age of Plentiful Energy? You may have heard about the scientist who gleefully reported that he had synthesized a universal solvent, a liquid that dissolves all known substances. Said his colleague, 'And in what kind of bottle will you keep it?'

We are looking at the nuclear counterpart of the bottle problem. In what kind of container can we heat plasma to 100 million degrees Celsius without destroying the container itself? Even the element with the highest melting point, tungsten, melts at a mere 3410°C and vaporizes at 5660°C. Scientists and engineers are working on a different approach: don't let the superheated plasma touch the sides of the container. This would be done by surrounding the container with a strong magnetic field that repels the plasma (which responds to magnetism) and forces it away from the sides into the centre. There, hopefully, it can attain fusion temperature long enough to be a steady source of energy. Fusion has in fact been achieved in this way numbers of times in experimental fusion reactors for a fraction of a second. These experiments are currently being carried out in various laboratories. Look for reports bearing such key terms (besides *fusion*) as *magnetic bottle, tokamak, stellarator, stabilized mirror,* and *laser-induced fusion*.

Three of these types — the magnetic bottle, tokamak, and stellarator — operate on the same principle, the magnetic containment (bottling) and compression of plasma. Therefore, we shall attempt to kill three birds with one stone: a brief description of the tokamak, designed by the Soviet physicist Lev Artsimovich, who in 1968 demonstrated his 'toroidal magnetic chamber' (for which *tokamak* is an acronym — in Russian). A toroid in edible form is called a doughnut. Thus a tokamak is a toroid-shaped hollow chamber operating on magnetic principles.

Essentially, a tokamak is a group of transformers with adaptations for its special functions. In each transformer, electric current, flowing through a primary coil (1), induces a magnetic field in the iron core (2); this in turn induces a current in the secondary coil (3). This coil is wrapped around a hollow toroid tube (4) containing fuel (deuterium and tritium). The secondary current sets up a magnetic field inside the tube, which in turn induces a current and a consequent magnetic field in the fuel itself (5). These opposing magnetic fields, around the secondary coil and around the fuel, repel each other, compressing the plasma from all sides. Meanwhile, the extremely dense flow of current within the plasma heats it up to enormous temperatures. The two forces — pressure and heat — hurl the plasma particles (deuterium and tritium) together with sufficient power to initiate fusion, forming helium and some by-products and releasing great quantities of energy, mainly in the form of heat. The heat is transported by a fluid to a conventional steam-driven turbogenerator for generating electricity.

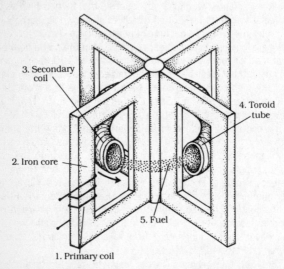

To end on a doleful note, there is indeed one successful way of accomplishing fusion — if you don't care about the bottle — or about humanity. A hydrogen bomb is literally a one-shot nuclear fusion reactor. It consists of two main parts: (1) a source of deuterium and tritium and (2) a heating device to achieve 100 million degrees Celsius. This heating device is another bomb, or several, of the old-fashioned, low-destructive-power fission type, the kind that was dropped over Japan and killed a mere 100,000 people.

See also COSMOLOGY; GRAND UNIFIED THEORY; NUCLEAR REACTOR.

NUCLEAR FISSION See NUCLEAR ENERGY

NUCLEAR FORCES

Called the strong force or interaction and the weak force or interaction, they are limited in influence to the atomic nucleus. They are two of the four known fundamental forces of the universe.

See also GRAND UNIFIED THEORY.

NUCLEAR FUSION See NUCLEAR ENERGY

NUCLEAR MAGNETIC RESONANCE See ANALYSIS

NUCLEAR-POWERED LASER See STRATEGIC DEFENSE INITIATIVE

NUCLEAR REACTOR

A device for releasing nuclear energy under *continuous, controlled* conditions (and therefore not a bomb). In some ways a common (fission-type) nuclear reactor is similar to a coal-burning furnace; here is a brief overview of the similarities:

In both cases, the principal output energy, heat, is converted to a more convenient form of energy, electricity. For example, in an electric power plant the heat converts water into high-pressure steam, to drive a steam turbine, which in turn drives an electric generator, whose output is sent through wires, far and wide. So a conventional power plant is quite similar to a nuclear power plant, most of the way.

	Coal Furnace	Nuclear Reactor
Fuel	Coal	Nuclear fuel
Action	Burning	Fission (atoms of fuel are split)
Output energy	Heat (and light)	Heat (and several forms of radiation)
Waste products	Ashes, smoke, combustion gases	Fission fragments, mostly radioactive

NUCLEAR REACTOR

The big difference is in the source of heat, inside the thick-walled, usually dome-topped building containing the nuclear reactor. This is where the nuclear fuel is stored and 'burnt'.

The common nuclear fuel is uranium 235. When U-235 atoms split (undergo fission), some fission fragments remain, and some matter and energy are released (expelled). The matter is mainly neutrons; the energy, mainly heat. This happens in nature, too, at a slow rate, an atom here and an atom there over thousands of years. But when the atoms are close enough and of a sufficient quantity (*critical mass*), their released neutrons accelerate the fission:

- Moderate acceleration: useful amounts of heat for running the steam turbine are generated.
- Greater acceleration: danger! More heat is generated than used; the excess heat would cause a rise in temperature to *meltdown* (about 1130°C). The highly corrosive, dangerously radioactive U-235 would destroy its containers and turn into a monster, never to be recaptured.
- Still greater (instantaneous) acceleration: a nuclear bomb – DANGER! a millionfold.

To keep the whole process going at a moderately humane rate, a moderator is employed. One type consists of *control rods,* made of neutron-absorbing material such as graphite or beryllium. These are interspersed among the fuel rods (long tubes filled with fuel pellets). Another type of moderator uses water circulating around the fuel rods and out to the next stage of operation.

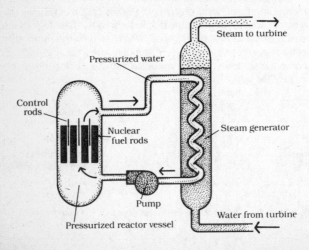

At the next stage the heated water is ready to do its work. In a *boiling-water reactor,* it boils into steam, which is piped to the steam turbine. In a *pressurized-water reactor,* the water is kept under pressure to prevent it from boiling whilst allowing it to reach a very high temperature. Its heat is transferred, via coiled pipes, to other water, which turns to steam for the turbine, and so on. In the *gas-cooled reactor,* the heat of fission is transferred to a gas such as helium or carbon dioxide. The heated gas is piped through a coil surrounded by water. The water turns to steam for the turbine.

Breeder reactors Is it possible to get something for nothing? Decide for yourself:

1. Most nuclear reactors use U-235 as their fuel, but the supply is limited.
2. Uranium coming out of the mines contains about 0.7% U-235 and about 99.3% U-238, which is not usable as a fuel but which can be made so (the technical adjective is *fertile*).
3. When U-235 undergoes fission, it gives off neutrons.
4. If U-235 is surrounded by a jacket of U-238, the neutrons convert the U-238 into *plutonium,* an artificial element that *is* a powerful nuclear fuel.

In a *breeder reactor,* the fuel rods are kept in use for about a year, during which time the U-235 gradually gives up its heat energy, as in a conventional reactor; meanwhile, the neutrons convert the surrounding U-238 into plutonium. Then the rods are lifted out and the spent U-235 is separated from the plutonium. Eureka! You have extracted energy from the U-235 *and* gained fuel — more fuel, in fact, than you began with. The difference is the something you got for nothing.

The breeder reactor sounds like the answer to our energy prayers, but like the whole nuclear energy field, it is riddled at present with unsolvable problems. Plutonium is highly toxic, and it is the basic 'explosive' for fission bombs; the theft of even a small amount of it by terrorists could pose a serious threat.

NUCLEI, UNSTABLE   See RADIOACTIVITY

NUCLEIC ACID   See DNA AND RNA

NUCLEON   See NEUTRON

NUCLEOTIDE   See DNA AND RNA

NUCLEUS, ATOMIC
See ELEMENTARY PARTICLES; NUCLEAR ENERGY

NUCLEUS OF A CELL See DNA AND RNA

NUCLEUS OF A COMET See SOLAR SYSTEM

OBSIDIAN See ROCK CLASSIFICATION

OCCLUDED FRONT See AIR MASS ANALYSIS

OCCULTATION

The passage of one astronomical object in front of another, obscuring it, as, for example, when the moon moves between the earth and a star, hiding it from view. The star may be thousands of times bigger than the moon, but the moon's relative nearness makes its apparent diameter much greater than that of the distant star. An analogy: a coin, held near your eye can occult a skyscraper a kilometre away. The best-known occultation (the term is from Latin *occultare*, to conceal from view) is the one in which the moon gets in front of the sun, blocking its light, which we call a *solar eclipse*.

One body passing in front of another doesn't always produce an occultation. Consider what happens every few years when the planet Mercury gets between the earth and the sun. The sun is relatively near to us, so its apparent diameter is large, and we perceive no occultation. Instead, Mercury is seen as a tiny dot moving across the face of the sun, an event called a *transit* of Mercury. Try to occult a skyscraper while standing across the street from it and holding a coin at arm's length. All you get is the tiny coin against the background of the vast skyscraper wall.

Many basic facts of astronomy were learned with the help of occultations. For example, the slow fading of a star's light as a planet passes in front of it reveals the existence of an atmosphere around the planet. Without the planetary atmosphere to diffuse it, the starlight would suddenly blink off. The sizes of many stars and planetary satellites and the existence of rings around the planet Uranus were also learned through observation of occultations.

OCEANOGRAPHY See PHYSICS

OESTROGENS See STEROIDS

OHM See ELECTRICAL UNITS

OIL SANDS See SYNTHETIC FUELS

OIL SHALE See SYNTHETIC FUELS

ONCOGENE See CANCER

ONCOLOGY

The branch of medical science, named from the Greek *onkos*, 'mass, bulk' dealing with abnormal growths, called *tumours*.
See also CANCER.

OORT COMET CLOUD See SOLAR SYSTEM

OPEN-HEART SURGERY

Open-*chest* surgery would be a more accurate name for these operations. They are performed for a number of disorders:

- To repair an ANEURYSM of the aorta (a weakened section of the artery).
- To correct loose, tight, sticky, or leaky valves in the heart, openings in the *septum* (the thick wall between the right and left sides of the heart), or transposition of narrowness of blood vessels in the heart.
- To widen or bypass clogged *coronary arteries*. In coronary bypass surgery, a section of a vein taken from the leg is put in place to shunt blood around the clogged part of the artery.
- To provide a temporary pump (for hours or days) to assist an ailing heart in its work of circulating blood throughout the body.
- To replace a hopelessly diseased heart with another human heart (a *heart transplant*) or with an artificial heart.

An operation that takes only a few minutes — for example, separating the parts of a sticky valve — can sometimes be done while the heart continues its beating. But complicated operations may take many hours. The heart must be stopped (and opened in some cases) for the operation. The cells of the body, especially those of the brain, can live only minutes without circulating blood, so these operations had to wait for the development of the *pump oxygenator*. This device is known popularly and accurately as the *heart-lung machine* because it takes the place of those organs. The oxygenator part, like the real lungs, removes waste carbon dioxide from the blood and adds oxygen. The pump, connected to two large blood vessels in the body, circulates the blood.

OPEN UNIVERSE See COSMOLOGY

OPERATORS (DNA) See DNA AND RNA

OPPORTUNISTIC DISEASES See AIDS

OPTICAL MICROSCOPE See MICROSCOPE

ORAL CONTRACEPTIVES See STEROIDS

ORGANIC CHEMISTRY
See CHEMISTRY; MOLECULAR BIOLOGY

ORGANIC EVOLUTION See EVOLUTION

ORGANIC MATTER See PROTEINS AND LIFE

ORIGIN OF SPECIES See EVOLUTION

OSCILLATING UNIVERSE

The hypothesis (also called the *pulsating* universe) that gravitation will cause the expanding universe to slow down and reverse, so that its parts come together again, prior to another big bang and cycle of expansion. If there is insufficient matter for this to happen, the universe will continue its expansion (open universe).
See also COSMOLOGY.

OSCILLOSCOPE

An apparatus similar to a television tube used mainly for graphing oscillating forms of energy, such as alternating currents, brain waves, and radio waves.

OSTEOARTHRITIS See ARTHRITIS

OSTEOPOROSIS See STEROIDS

OUTPUT See COMPUTER

OVERTONES See SOUND RECORDING; TONE CONTROL

OXIDATION See ENERGY

OZONE See OZONE LAYER

OZONE LAYER

A band of gas in the atmosphere, about 15 to 50 kilometres (12 to 30 miles) above the surface of the earth; also called the *ozonosphere*. The gas — *ozone* — is a form of oxygen containing 3 atoms in each molecule (O_3), which is formed by the action of solar ultraviolet radiation on ordinary oxygen (O_2) in the upper atmosphere.

Near ground level, ozone is an undesirable pollutant, a constituent of smog that irritates the eyes and impairs breathing. But in the upper atmosphere it acts as a screen against harmful ultraviolet rays which would otherwise reach earth, causing extreme sunburn, skin cancer, and irreparable damage to the body's proteins and nucleic acids (DNA and RNA).

Scientists monitoring the ozone layer had noticed that it became thinner each Spring over the South Pole, and in 1985 researchers from the British Antarctic Survey discovered a hole in it. The major cause of this is the breakdown of ozone by chlorine from compounds called *chlorofluorocarbons (CFCs)*. These substances, which can remain in the atmosphere for up to 100 years, are used extensively as refrigerants, in air-conditioning systems, in foamed plastics, and (though less frequently now) as aerosol propellants.

By late 1988 fear of further damage to the ozone layer had the Environmental Protection Agency of the United States to call for a complete ban on the use of CFCs. Manufacturers had already found substitutes for CFCs in aerosols and were competing to find the first alternative to use in refrigerators and cooling systems.

OZONOSPHERE See OZONE LAYER

PACEMAKER, CARDIAC See CARDIAC PACEMAKER

PALAEONTOLOGY See BIOLOGY

PARACENTESIS See AMNIOCENTESIS

PARALLEL (OF LATITUDE)
See CELESTIAL COORDINATES

PARATHYROID GLAND See HORMONE

PARATHYROID HORMONE See HORMONE

PARKINSON'S DISEASE

A disorder of the nervous system, resulting in gradual loss of muscular control. In 1817 James Parkinson (1755–1824), an English physician, described some of its more prominent symptoms: tremors of the hands, arms, or legs; slow and stiff movements; difficulty in walking; stooped posture; and a fixed, staring facial expression. These symptoms may follow brain injury, poisoning, or certain diseases, but in most cases the basic cause of Parkinson's is not known. The individuals affected are usually middle-aged or elderly. About 15 in every 10,000 people in the United Kingdom are victims of Parkinson's disease.

No cure is known, but several drugs, particularly *levodopa (L-dopa)*, are used to lessen the severity of the symptoms. L-dopa is changed by the brain to *dopamine*, a substance deficient in the brain cells of Parkinson's patients. Dopamine is one of a group of substances, called *neurotransmitters*, that are involved in the generation and sending of nerve impulses.

PARTIAL ECLIPSE See ECLIPSE

PARTICLE ACCELERATOR

Formerly called an *atom smasher*, but a more accurate name might be 'nucleus smasher'. These powerful machines are used to find out what atomic nuclei are made of by smashing them and examining the pieces. A student of clock repairing wouldn't learn much from such a form of mayhem, but it works very well for the student of nuclear structure, the physicist. The difference is that nuclear particles do not lie limp and mangled after liberation; they retain their individual lively forms of behaviour. Magnetic particles continue to be affected by magnets, electrically charged particles continue to respond to electrical fields, and heavier particles continue to deliver stronger impacts than lighter ones, like footballs colliding with Ping-Pong balls. These subnuclear responses are measured and recorded mainly by instruments called *cloud chambers* and *bubble chambers*.

Particle accelerators have enabled scientists to achieve partial answers to questions in pure research — What is matter? What is energy? How did the universe begin? — as well as spin-offs, both useful and destructive, in applied fields such as radiation therapy, nuclear electric energy, and the manufacture of nuclear bombs.

Particle accelerators come in a range of sizes, from 28 centimetres (11 inches) in diameter (E. O. Lawrence's 1929 *cyclotron*; Greek *kyklos*, circle) to more than 80 kilometres (50 miles) in circumference (the proposed Superconducting Super Collider). They work in various ways, but all do the same basic job: they accelerate particles. These particles are tiny 'hammers' that are hurled with precise aim at target nuclei the hammers are themselves nuclear particles: protons, neutrons and others (breaking rocks with hammers made of rock). The hammer particles are given energy by being speeded up from a standstill to several thousand kilometres per second when they strike their target.

There are numbers of ways to achieve such acceleration, in various designs of machines. All these designs involve one or more of the following principles:

The playground-swing principle On a playground swing it's possible to achieve great heights by repeated small additions of energy (pushes or pulls) properly timed. Particle accelerators deliver magnetic and electrical pushes and/or pulls to the hammer particles orbiting in circular chambers. When, in a cyclotron, or a more advanced form, a *synchrocyclotron*, particles achieve maximum speed, they are diverted from their circular path by electromagnets, so as to strike the target substance whose nuclei are to be split apart.

The slit-doughnut principle Cut one slit across an uncooked doughnut and you can straighten it into a rod. Similarly, the hollow ring-shaped form of a cyclotron can be straightened out into a tube. Electromagnetic 'kickers' surround the tube at precisely spaced intervals. The longer the tube, the more space for more kickers, and the higher maximum speed achieved. The accelerator at Fermilab near Chicago is 3.2 kilometres (2 miles) long. Because the particles travel in a straight line, the machine is called a *linear accelerator*.

The colliding-beam principle A car going 80 kilometres (50 miles) an hour hits a thick concrete wall. The collision delivers a certain amount of punishment to the wall, the car, and the driver. When two cars, each going 80 kilometres per hour, collide head on, each car and driver feels the impact of hitting a wall at 160 kilometres (100 miles) an hour. In a colliding-beam accelerator, two separate sets of particles are accelerated in opposite directions around a circular track, precisely steered by electromagnets. As they achieve maximum speed, they are finally steered into a head-on collision.

The opposites-attract principle Suppose, in the aforementioned case of vehicular homicide, each car had been equipped at the front with a powerful magnet, with opposite poles facing — each one's south pole facing the other's north pole. That added attraction would contribute to an even more violent smash. Similarly, some accelerator experiments use opposite-attracting particles, such as protons and antiprotons, for a super-sized collision.

The relativity principle This one is hard to believe but will become a bit more credible as you read the entry on relativity. An object gaining speed also gains mass. The greater the ultimate speed, the greater the gain in mass. At top speed, a particle may be as much as 40,000 times as heavy as when it began its trip. But that gain isn't entirely free: a heavier hammer requires more force to lift it, and heavier particles require more electromagnetic force to accelerate them. But the result — which is what really matters — is quite smashing.

In the forthcoming years, larger and more powerful accelerators will be built, and perhaps new principles of accelerator design will be discovered. The news media will carry many items on the subject. Here are some important terms you will come across:

Electron volt The unit by which accelerator output is measured (*see* ELECTRON VOLT).

CERN The *Conseil Européen pour la Recherche Nucléaire* (European Organization for Nuclear Research, now called the European Laboratory for Particle Physics) is a consortium of 12 European nations, based in Geneva, Switzerland. CERN has probably taken the lead in nuclear research by adhering to the principle that in union there is strength, as in the laser system of coherent application of energy.

The Superconducting Super Collider (SSC) In 1987 Ronald Reagan, the president of the United States, approved the building of an accelerator more than 80 kilometres (50 miles) in circumference. Within this colossus thousands of huge magnets will accelerate two beams of protons to energies of 20 trillion electron volts each and steer them into head-on collisions. The SSC will be the most powerful accelerator in the world — if CERN's next project hasn't overtaken it.

The matter of 'most powerful' deserves mention. It isn't simply a case of 'more is better' but that more power enables scientists to split farther and farther down the scale of ELEMENTARY PARTICLES. The farthest down anyone has yet gone is the quark, but more powerful accelerators may split the quark, too.

Storage rings Doughnut-shaped hollow chambers where particles such as electrons and positrons are kept waiting, behind the scenes, racing around at moderate velocity, until needed on stage for the main performance.

Supercooling A particle accelerator's power depends mainly on the strength of its electromagnets; this in turn depends mainly on the strength of the current flowing through its wires — and here we come to two roadblocks:

1. Electric current causes a rise in temperature; the stronger the current, the higher the rise.
2. A rise in temperature causes a rise in resistance; the electromagnet becomes a poorer conductor of electricity — a kind of self-limiting impasse. This is where supercooling will be making the news; in ways of cooling the electromagnet coils down to a temperature near ABSOLUTE ZERO.

See also SUPERCONDUCTIVITY.

PARTICLE BEAMS *See* STRATEGIC DEFENSE INITIATIVE

PARTICLE PHYSICS *See* PHYSICS

PARTICLES, ELEMENTARY
See ELEMENTARY PARTICLES

PASCAL *See* COMPUTER LANGUAGES

PASSIVE SMOKING *See* SMOKING

PASSIVE SONAR *See* SONAR

PATHOLOGY *See* MOLECULAR BIOLOGY

PCBs

Substances (*p*olychlorinated *b*iphenyls) that were once widely used in inks and paints and in oils as insulating fluids in electrical TRANSFORMERS. PCBs were banned in the United States, the major producer, in 1977 after it was found that they were linked to birth defects, serious liver disorders, and cancer. Despite an international ban on their manufacture from the early 1980s, enormous quantities of PCBs still exist, leaking gradually into soil and water where they remain, virtually impossible to get rid of. So widespread are they that scientists believe it possible that the entire populations of the United States and Britain contain PCBs in measurable amounts. Animals such as seals are particularly vulnerable because they eat contaminated fish that has fed on contaminated plankton. The chemicals become concentrated in the seal's body fat and may lower resistance to disease.

A continuing environmental problem is the danger of spills or fire in the hundreds of thousands of oil-filled electrical transformers still in use. 5 tonnes of PCBs contained in such transformers were dumped into the North Sea when the Piper Alpha oil platform exploded in 1988. In a fire the PCBs may give off DIOXINS, one of which is among the most toxic substances known.

PENICILLIN *See* ANTIBIOTICS

PENUMBRA *See* ECLIPSE

PEPTIDE BOND *See* PROTEINS AND LIFE

PERIPHERALS

Computers are sometimes glibly likened to human brains. To play with that parallel for a moment, peripherals could then be regarded as the sensory organs that transmit outside information (input) to the brain, or as other organs, mainly muscles and glands, that receive the brain's commands and obey them (output). Some input devices include a typewriterlike keyboard and a bar code scanner at a supermarket checkout. Some output devices are a monitor or screen and a printer.

Still playing with that parallel, inputs to humans, of sight, sound, smell, touch, and taste, must be converted into a single kind of message to the brain: chemical-electrical nerve impulses. A place where such a conversion occurs is called an interface. Likewise,

PHARMACOLOGY See MOLECULAR BIOLOGY

PHENYLKETONURIA (PKU)

A hereditary disorder leading to mental retardation if untreated.
See also ae3 GENETIC DISEASES

PHOSPHOR See CATHODE RAY TUBE

PHOTINO See GRAND UNIFIED THEORY

PHOTON

A particle that transmits the energy in light, X-rays, and other forms of electromagnetic radiation.
See also GRAND UNIFIED THEORY; RADIANT ENERGY.

PHOTOSPHERE See SOLAR SYSTEM

PHOTOSYNTHESIS See ENERGY

PHOTOVOLTAIC SYSTEM

A means of converting radiation (especially light) directly into electricity.
See also ENERGY RESOURCES.

PHYSICAL CHEMISTRY See CHEMISTRY

PHYSICS

there are interfaces wherever brain messages are converted into instructions to muscles and glands. In computers there are electronic interfaces at every place where a COMPUTER is joined to a peripheral.

'The quantitative branch of science dealing with the nature of matter and energy and the relationship between them.' This standard definition gives the impression that physics covers the entire universe — as indeed it does. In a sense,'If it's measurable, it belongs to physics'. This excludes such intangibles as religion, free will, and the emotions (in spite of 'How do I love thee? Let me count the ways') but includes almostt everything whose tangible essence is measurable.

Physics in the 17th century was called natural philosophy and involved the study of all aspects of the material world, animate and inanimate. The animate aspect separated out as biology, and part of the inanimate became chemistry.

Physics may be divided into several branches:

Astrophysics The physics of the 'big stuff': the big bang, galaxies, solar systems, white dwarfs, red giants, and black holes, among others; what they are, how they work, how they came to be, what may happen to them and why. In order to understand the dynamics of these huge entities, it is necessary to understand the physics of their very smallest components: protons, photons, electrons, muons, and other members of a large tribe of particles.

Biophysics The physical aspects of living systems. For example, what mechanism lifts water to the topmost branches of a 100-metre (330-foot) sequoia? What are the electrical actions in nerve conduction? What human engineering requirements must be fulfilled in order to design an artificial heart?

Geophysics The study of the planet earth as a set of physical systems. What produces ocean tides? What forces cause the formation and storage of petroleum in certain places and not in others? What is the source of the earth's inner heat? Some of the principal areas of geophysics are seismology (study of earthquakes and other movements of the earth's crust), meteorology (study of the earth's atmosphere), and oceanography (study of the ocean composition and movement).

Particle physics The study of the structure of matter at its smallest scale, its most elementary particles, which cannot be further subdivided. This study requires the use of a PARTICLE ACCELERATOR operating at very high electrical energy. Also called *elementary particle physics* and *high-energy physics*.

Plasma physics A plasma is a gas raised to a temperature so high that its atoms have been shaken apart into their constituent electrons and nuclei. In such a state the nuclei, under extreme pressure, can be forced to combine with each other to form larger nuclei. In the process, enormous quantities of energy are released. This is the basis of energy generation by fusion.

Quantum physics Also called *modern physics* based on the quantum theory of energy, stated in 1900 by Max Planck. Prior to Planck, energy was regarded as being infinitely divisible, just as a centimetre can be infinitely divided into smaller units. Similarly, a candela of light, a volt of electrical pressure, or a joule of heat could be infinitely divided. Planck showed that there is in each case a limit of smallness, a quantum (Latin *quantus*, how much?) of energy, describable in an equation later named Planck's law. Quanta are obviously very small, but the consequences of their existence are very great.

Solid-state physics Also called *condensed-matter physics*; the study of solid substances in relation to their physical properties:

Mechanical Steel wire has high tensile strength and elasticity. Lead is almost totally lacking in these properties. How is this explained by their molecular structure?

Electrical Metals are good to excellent conductors of electricity. Plastics and ceramics are poor conductors (they are insulators). Certain elements, crystalline solids such as silicon and gallium, are in-between and as such are called *semiconductors*. These varying properties are the major basis of the entire electronics technology. What determines electrical conductivity?

Magnetic Cupboard door latches work with permanent magnets made of steel. Doorbells work with temporary magnets made of steel. How do they differ?

Optical Two camera lenses of exactly the same size and shape can have entirely different focusing powers. How?

Thermal Good conductors of heat are also good conductors of electricity. What molecular likeness explains this?

In almost all solids, the molecules are arranged in a regular, repeating, crystalline form (e.g., quartz, sand, salt, sugar). The study of the properties of crystals is a branch-within-a-branch of solid-state physics called *crystallography*.

See also ELEMENTARY PARTICLES; NUCLEAR ENERGY.

PHYSIOLOGY See BIOLOGY; MOLECULAR BIOLOGY

PIEZOELECTRIC EFFECT See ULTRASONICS

PION See ELEMENTARY PARTICLES

PITUITARY GLAND See HORMONE

PIXEL See CATHODE RAY TUBE; VIDEOTAPE

PKU See PHENYLKETONURIA

PLACEBO   See SCIENTIFIC TERMS

PLANET   See SOLAR SYSTEM

PLANETOID   See SOLAR SYSTEM

PLASMA (ATOMIC)   See NUCLEAR ENERGY

PLASMA PHYSICS   See PHYSICS

PLASMID   See GENETIC ENGINEERING

PLASMID CHIMERA   See GENETIC ENGINEERING

PLATE TECTONICS

A theory that explains how certain forces have shaped the major landmasses and oceans of the earth. *Tectonics* derives from the Greek *tekton*, 'carpenter, builder'. A more suitable term might be developed from the Latin *coquus*, 'cook', or the Old English *bacan*, 'to bake'.

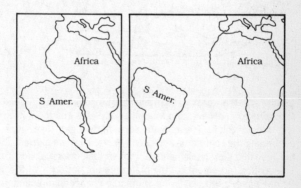

Consider this map. The shapes are familiar, but the contiguity is not. Africa and South America seem to be adjoining pieces in a giant picture puzzle. These two continents were part of a huge landmass that split into several pieces about 160 million years ago. South America (as yet unnamed, needless to say) drifted away, a fraction of an inch per year, to where it is today, on the earth and in your atlas. Nor is this the end of the *continental drift*: someday your atlas will be obsolete. The verb *drift* implies an object (something solid, having a shape) being carried along on a fluid (something soft and shapeless, in motion) — which is a fairly reasonable way to describe the basic idea of plate tectonics.

Picture a huge cauldron of thick porridge being cooked. Afloat on the porridge are many slices of toast. Heat from beneath causes the porridge to rise, spread out and sink, over and over, in the process called convection. The slices of toast, some of them bumping, are carried along by the horizontal part of the convection.

Our planet is like a giant cauldron, whose principal source of heat, deep within, is the radioactive decay of various elements. The heat causes much of the earth's interior to remain in a molten or soft-solid (plastic) state. Here and there, within the most fluid layers, convection currents circulate, like the up-horizontal-down flow of the cooking porridge.

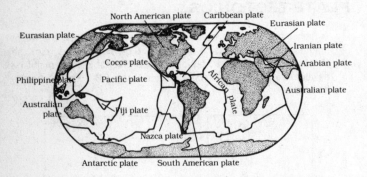

Now for the toast. The upper part of the earth's crust (the *lithosphere*) is divided into a number of plates about 95 kilometres (60 miles) thick. These plates are fairly rigid, like huge rafts containing the continents and ocean floors. They are carried along by the (usually) slow movements of molten rock (*magma*) underneath. The drifting movements of the plates are not uniform; a little slower for one plate, a little faster for another, a change in direction, a halt, a collision. The result is one or more of these reactions, among others:

- One plate may slide past another. Their displaced edges form a *transform* fault, as in California's well-known San Andreas system, extending about 1000 kilometres (600 miles) diagonally down the state. The edges, not being straight, smooth, or lubricated, move in a chattering action called *strike-slip* motion. (Press the edges of your hands together, *hard*, and try to push one hand forward.) The strike part is the motionless forward pressure, building up energy; the slip is the sudden release of the pent-up energy — an *earthquake*. Along the chattering edge, a weak spot may develop, and magma may force its way up and out — a *volcano*.

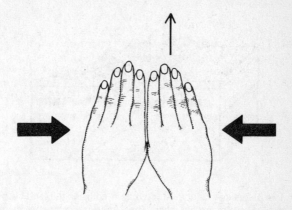

- Two plates may drift apart, leaving a lowered groove, a *rift valley*. If this occurs in the depths of the ocean, it is called a *seafloor spreading*. In the stretched, thinned seafloor a break may occur and fresh magma may well up and form a mountain range. There is just such an underwater mountain range over 64,400 kilometres (40,000 miles) long snaking its way through the North and South Atlantic, Indian, and South Pacific oceans.
- A plate may split beneath a continent; the land above the split subsides — again a rift valley. The Great Rift Valley and its connecting valleys extend over 4830 kilometres (3000 miles) through East Africa and the Middle East. They may have been formed by such a split.

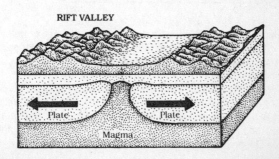

PLATE TECTONICS

- One plate may prize its way (strike-slip again) under the edge of another. Its advancing edge bends downwards in a process called *subduction* (from Latin words meaning 'to lead under'). It reaches hotter zones beneath, melts, and becomes part of the magma.

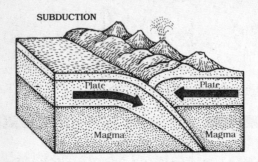

- Two plates may collide. At the zone of collision the land heaps up, producing a mountain range. The Himalayan ranges are the result of India pressing northwards against the Asian mainland.

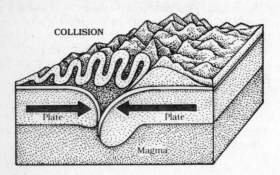

Most tectonic behaviour is slow, in the time range of millions and hundreds of millions of years. The few spectacular exceptions are the slip of strike-slip actions (earthquakes) and breakthroughs such as volcanoes.

One of the most dramatic examples of plate tectonics is displayed along the plates that include the Pacific Ocean. The contact zones between plates are riddled with earthquakes, studded with volcanoes, and ringed by mountain ranges. This region is aptly named the Ring of Fire.

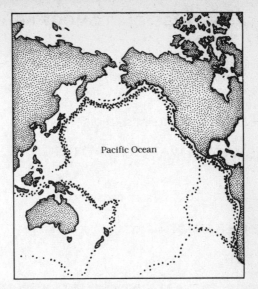

PLUTONIUM See NUCLEAR REACTOR

PNEUMOCYSTIS PNEUMONIA See AIDS

POINT PLOTTING See COMPUTER GRAPHICS

POLAR CONTINENTAL AND MARITIME AIR MASSES See AIR MASS ANALYSIS

POLICE RADAR See DOPPLER EFFECT

POLYCHLORINATED BIPHENYLS See PCBs

POLYMER See CHEMISTRY

POLYMER CHEMISTRY See CHEMISTRY

POLYUNSATURATED FATS See CHOLESTEROL

POSITRON

A subatomic particle with characteristics identical to those of the electron, except for its positive (+) charge; electrons are negatively (−) charged.
See also ANTIMATTER.

POSITRON EMISSION TOMOGRAPHY (PET)

A scanning technique similar to the CAT scan (see COMPUTERIZED AXIAL TOMOGRAPHY but using positrons in place of X-rays.
See also ANTIMATTER.

POTENTIAL ENERGY *See* ENERGY

PRESSURIZED-WATER REACTOR
See NUCLEAR REACTOR

PRIMARY ROCK *See* ROCK CLASSIFICATION

PRIME MERIDIAN *See* CELESTIAL COORDINATES

PRINTER

A computer-driven electric typewriter, whose output reasonably enough, is called a *printout*. The earliest printers were actually electric typewriters adapted to computer operation. Modern printers are mainly of four types:

Dot-matrix printers The characters are created from a basic group, or *matrix*, of dots. The matrix is composed of tiny pins, with electromagnets behind them. A computer signal to an electromagnet causes its pin to move forward, striking a ribbon and printing a dot on paper. Shown here is the letter e and its enlargement to show detail. Dot-matrix characters are not beautiful, but they are legible and the printer is speedy and relatively cheap.

Thermal printers These are similar to dot-matrix printers, except that a special heat-sensitive paper is used, no ribbon is required, and the pins don't move. A computer signal to any pin causes it to heat up for a fraction of a second, forming a dot on the paper.

Daisy wheel printer The printing characters are at the end of plastic fingers that radiate like daisy petals or spokes of a wheel from a central hub. The wheel rotates on a shaft, directly behind a typewriter ribbon. An electromagnet, when signalled by a computer, causes a tiny hammer to strike the end of a spoke against the ribbon, thus printing a character on the paper. In use, the wheel spins constantly, so that the characters are caught whilst in motion, about 20 per second.

Laser printers These are nonimpact machines: that is, there is no piece of type striking a sheet of paper. Instead a computer controls the movement of a very thin beam of LASER light so that it forms letters on a sheet that is then pressed against a sheet of paper, as in a photocopying machine.

PRINTOUT   See PRINTER

PROBENECID   See STEROIDS

PROGESTERONE   See STEROIDS

PROMOTERS (DNA)   See DNA AND RNA

PROPELLER

A fanlike device turned by an engine. Turning in water or air, the propeller, as befits its name, propels the boat or aeroplane to which it is attached. Two kinds of action are involved:

1. The propeller blades are set at an angle; as they rotate, they push air or water backwards and thus simultaneously thrust the aeroplane or boat forwards (see NEWTON'S LAWS OF MOTION).
2. The blades are shaped in a special way that reduces the pressure in front of them, thus adding to the forward thrust.
   See also AERODYNAMICS.

PROTACTINIUM   See RADIOACTIVITY

PROTEINS AND LIFE

Until the early 19th century, living things were regarded as being composed of a special class of substances, *organic matter*, endowed with a unique life force, *élan vital*, which inhered not only in the living organism in action but even in its excretions and castoffs. This theory was first assaulted by the German chemist Friedrich Wöhler (1800–1882), who in 1828 announced the results of a simple experiment. He had heated a totally nonliving (inorganic) salt, ammonium cyanate, and produced *urea*, the principal ingredient (other than water) of urine.

There were more assaults on the life-force theory, leading to the *coup de grâce*, but we have neither the space nor the inclination to beat a horse, dead or alive. Let us instead rush in and deliver a

PROTEINS AND LIFE

bare-bones description of living things that applies to all life on earth.

A living thing, or *organism*, is composed mainly of proteins whose production is controlled by nucleic acids (DNA AND RNA). Even the most minute, simplest organisms that dare to count themselves among the marginally living, the VIRUSES, are protein; a virus organism consists of a collection of DNA or RNA particles surrounded by a coat of protein. The virus shown in this illustration causes mosaic disease in tobacco plants.

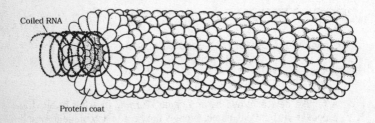

All the various proteins of all living organisms consist almost entirely of four elements; carbon, hydrogen, oxygen, nitrogen (CHON, easy to memorize). Proteins are made of *amino acids*, and all amino acids contain these four elements, combined in different ways.

GLYCINE

$$H-\underset{\underset{H}{|}}{\overset{\overset{H}{|}}{N}}-\underset{}{\overset{}{C}}-\overset{\overset{O}{\|}}{C}-OH$$

Amino 'head'   Acid 'tail'

On the next page is the simplest amino acid, *glycine*. As you can see, it's just a lot of CHON hooked up in a peculiar way. The CHON stuff in glycine could have been written simply: $C_2H_5O_2N$. But this diagrammatic way, called a *structural formula*, reveals more: it shows how the various atoms are joined to each other. The $H-\overset{\overset{H}{|}}{N}$ is the *amino* part of the molecule — the 'head', if you will. The $\overset{\overset{O}{\|}}{C}-OH$ is the *acid* part, or 'tail'.

PROTEINS AND LIFE

Here is the structural formula for the amino acid called *alanine*. Again, a lot of CHON. Notice that the central part within the dotted line is exactly the same as in glycine. But the outside part, called the *side chain*, is different.

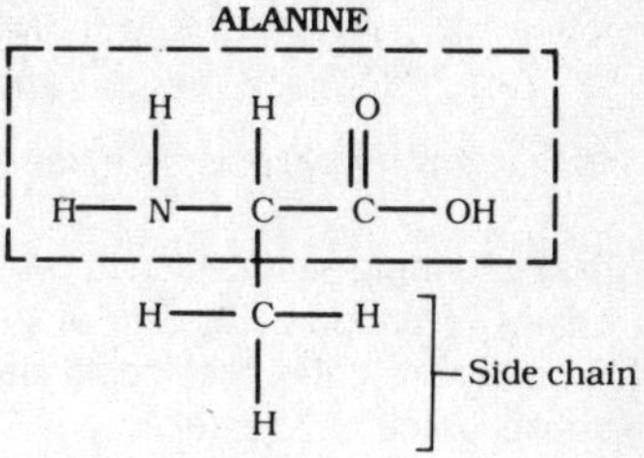

ALANINE

```
        H   H   O
        |   |   ||
    H—N—C—C—OH

        H—C—H    ⎤
            |     ⎬ Side chain
            H    ⎦
```

Amino acids are all alike in their central section and all different in their side chains. Scientists believe that this clustering of elements, forming CHONs, happened billions of years ago, before life began. There was plenty of carbon, hydrogen, oxygen, and nitrogen around and plenty of energy, perhaps in the form of lightning bolts, to slam the atoms together into compounds.

Amino acids combine to form proteins in a simple, ingenious way. Behold!

Slide the two amino acids towards each other. The two ends approaching are H and OH, which feel a strong attraction towards each other (never mind why). They join and form HOH, commonly written as H_2O – water. That urge to join into water frees the amino end of one molecule. It links to the newly freed acid end of the next, in a *peptide bond*. Like the trunk-to-tail-to-trunk formation of elephants in a circus parade, more such bonds form, to produce a chain of amino acids virtually as long as you wish. And there, at last, long ago, it happened: the giant step that transformed inert matter, the amino acids, into something with the potential for life, the proteins.

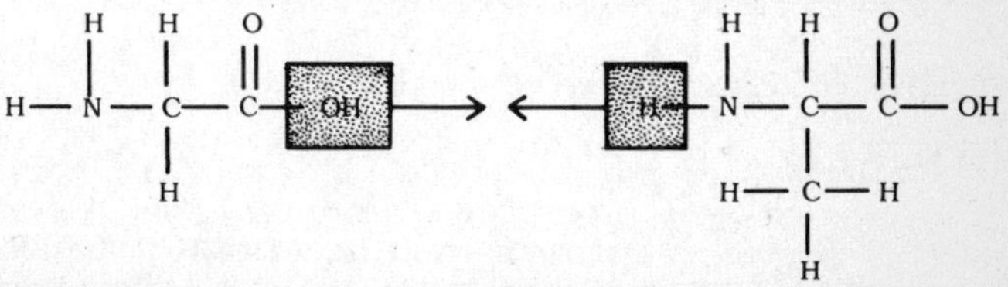

231

PROTEIN SYNTHESIS

```
    H   H   O                    H   H   O
    |   |   ||                   |   |   ||
H — N — C — C ━━━━━━━━━━━━  N — C — C — OH
        |      Peptide bond      |
        H                    H — C — H
                                 |
                                 H
```

Three major (and many minor) reasons made the difference:

1. The potential for forming peptide bonds permitted the building of very long chains of amino acids, allowing the formation of all kinds of complex molecules that could become shaped structures: threadlike, granular, sheetlike.
2. An organism is alive only as long as it continues to carry on thousands of coordinated chemical and physical reactions. Some large, complex proteins called *enzymes* are intimately involved in these reactions.
3. The large number of amino acids permitted the building of many kinds of proteins. Suppose there were only five different amino acids, called, for convenience, 1, 2, 3, 4, and 5. You could build a string 1, 3, 5, 2, 4, or 5, 2, 4, 3, and 1 — altogether, 120 possible arrangements. Add a sixth amino acid and you get 720 arrangements. Add a seventh and get 5040. In fact there are more than 20 kinds of amino acids available, each of which can be used many times. The result is millions of possible large, heavy proteins, to form bone tissue, cartilage, skin, haemoglobin, insulin, and on and on.

What determines which amino acids are to be selected and arranged into proteins? Read the entries DNA AND RNA, ENZYME, and VIRUSES.

PROTEIN SYNTHESIS See DNA AND RNA

PROTON See ATOM; ELEMENTARY PARTICLES; NUCLEAR ENERGY

PROTOSTAR See STELLAR EVOLUTION

PSI PARTICLE See ELEMENTARY PARTICLES

PULSAR

A celestial object that emits on/off radio signals (pulses) with remarkable regularity. Pulsars were discovered by Antony Hewish and Jocelyn Bell at Cambridge University in 1967, using radio telescopes. Evidence collected since that time makes it almost certain that pulsars are *neutron stars*, rotating at very high speeds, up to 30 times per second. The pulses may arise from the high-speed motion of electrons in the pulsar's powerful magnetic field.

See also RADIO ASTRONOMY.

PULSATING UNIVERSE *See* OSCILLATING UNIVERSE

PULSE *See* CIRCULATORY SYSTEM

PUMP OXYGENATOR *See* OPEN-HEART SURGERY

PUNCH PRESS *See* MANUFACTURING PROCESSES

QUADRAPHONICS

A system of music recording that proves there can be too much of a good thing. Music was first reproduced by the *monophonic* system (Greek *monos*, single + *phon*, sound), in which both of the listener's ears heard the same sounds, coming from one loudspeaker. You could add a second speaker, but the right ear still heard the same sounds as the left.

Then came the *stereophonic* system, designed to produce a more solid, three-dimensional effect (Greek *stereos*, solid). The sound is recorded by two (or more) microphones, placed right and left in relation to a group of musicians. The resulting record or tape contains two soundtracks. The tracks are slightly different from one another because the left microphone, being closer to the musicians at the left, picks up their sound slightly sooner and louder than does the microphone at the right. The reverse situation occurs with the microphone at the right.

The two tracks are played separately through equipment that feeds two loudspeakers, left and right, separately. The sound from the right speaker is slightly different from the sound from the left speaker. Your ears hear slightly different sounds (as they do in ordinary 'real' hearing), and you are beguiled into thinking you are in the presence of a three-dimensional musical group rather than a point source of sound. Stereophonic sound is more convincing and realistic than monophonic.

If two are better than one, shouldn't four be better than two? The *quadraphonic* system employs four microphones placed at left front, right front, left rear, and right rear. The music is recorded on four tracks and played through four speakers placed at the four corners of the room. The resulting sound is stunningly realistic and awesomely complicated and expensive. It is now mainly a historical curiosity.

QUALITATIVE AND QUANTITATIVE ANALYSIS   See ANALYSIS

QUANTUM PHYSICS   See PHYSICS

QUANTUM THEORY

Nothing in your daily experience will lead you to believe in the quantum theory of Max Planck, German physicist (1858–1947). The theory deals with the energy in molecules, atoms, and their subparticles.

In your daily experience, this kind of thing happens: you stick a thermometer into a bowlful of hot water, and the line moves up smoothly along the scale, never skipping a number. Similarly, you step on the accelerator and you car speeds up: the speedometer shows a rise in miles per hour, smoothly, never skipping a number. Whether you speed up or put on the brakes, there is always a transition from point to point, no matter how close together you choose your points.

Now come down to the scale of molecules and atoms and meet two surprises. First, everything is in a state of endless motion — spinning, revolving, vibrating, at immense rates of speed, in the millions and thousands of millions per second. Second, these rates of motion, when they change, do so in abrupt jumps (*quantum* jumps), with no transitions between.

Why does the quantum theory operate in the molecular world and not in your daily experience? Because instruments such as thermometers and speedometers measure large-scale manifestations of *energy* changes. In the flame of a burning match, billions of molecules undergo chemical change. Looking at a flame is like viewing the moon with the naked eye. You see a disc with some markings. Look through telescopes with higher and higher magnifications, and you begin to see craters, rills, and chains of mountains. Looking at our common experiences — burning, heating, cooling, shining, booming, bubbling — through increasingly powerful and more precise instruments from microscopes to particle accelerators, we find that our common experiences are the vast sum total of energy changes in and between molecules and their parts: protons, neutrons, electrons, quarks, and scores of others.

All these energy changes are subject to Planck's quantum theory: when we change the energy content of these particles, they respond in quantum jumps. Lay an iron nail in a fire (adding heat energy), and some outer electrons in the iron atoms jump to farther-out orbits — exact orbits, not just a little more and still a little more. All the motions of particles — spin, revolution, vibration — change by quantum jumps.

Does the quantum theory have any practical value? Every device whose operation is based on energy changes within molecules must be designed with the theory in mind. The telephone, calculator, computer, microwave oven, radar, CAT scanner, and laser are only the beginning of a very long list.

QUARK

A particle that may be the fundamental unit of matter. There are six kinds (flavours, as they are called) of quarks, and each flavour has three varieties (called colours). Like all known particles, quarks have their antimatter opposites, known as antiquarks.

See also ELEMENTARY PARTICLES.

QUARTZ

A crystalline substance, silicon dioxide, the chief constituent of most kinds of sand; the most common of all minerals. Tiny quartz crystals, properly prepared, are used to control the timekeeping of watches.

A quartz crystal acts as a time divider, performing the same job done by pendulums or balance wheels in mechanical (windup) clocks and watches.

A pendulum is caused to swing (oscillate) by a wound-up spring. The oscillations control the movements of the hands. A quartz crystal is caused to oscillate (get longer and shorter, by an infinitesimal amount) by electric current from a tiny battery flowing through an electronic circuit. The crystal's oscillations control the movement of the hands of the timepiece. In another type of timepiece, the crystal controls the display of numbers. Many electronic devices such as radio transmitters and TV cameras are crystal-controlled.

Quartz timepieces, even the modestly priced ones, are very accurate, to a minute or two per year. A spring-driven timepiece of that accuracy costs several thousand pounds and is accurate only if kept at a constant temperature, in a fixed position. A quartz crystal's rate of oscillation is only slightly affected by changes in temperature or position. The common rate is 32,768 per second. This dizzying number is processed through a tiny electronic dividing machine, where it is divided by 2, over and over again, 15 times, down to a rate of one oscillation per second, which can be handled by the dial machinery.

See also DIGITAL AND ANALOGUE.

QUARTZ WATCHES *See* QUARTZ

QUASAR

Short for *qua*si-stell*ar*, meaning 'star-resembling'. Since 1960, about 5000 quasars have been located and observed through optical and radio telescopes and catalogued, but their actual physical identities have yet to be specified. Details gathered to date are tantalizing indeed:

1. Most quasars are extremely bright, the brightest objects in the universe. The light of a single quasar is 100 or more times as bright as an entire average-sized galaxy. However, quasars appear faint even in large optical telescopes because they are extremely far away, towards the limits of the known universe (up to 12,000 million or more light-years).
2. Most are racing away at enormous speeds, some as fast as 90% of the speed of light, as evidenced by the shift towards the red in their spectra (*See* DOPPLER EFFECT).
3. Most are 8.05×10^{11} kilometres (500,000 million miles) or more in diameter, over half a million times the diameter of our sun.
4. Most do not shine steadily; their brightness varies periodically, some over a day or less, some over several years.
5. Most are probably not stars, although they show up as pinpoints in even the most powerful telescopes. They may be small galaxies so far away as to look like pinpoints. In that case why not just assume they *are* galaxies until further notice? See point number 1. Their brightness, say astronomers, cannot be achieved by any known process but perhaps by an energy transformation still to be discovered.

RADAR

Acronym for *ra*dio *d*etecting *a*nd *r*anging, a technique and apparatus for determining the location of an object by the use of radio waves. The most visible and ubiquitous aspects of radar are the rotating, curved-surface antennas seen on top of most ships and airport towers. Not visible, but equally important, are the radar antennas hidden in the noses of aeroplanes.

Radar antenna

In operation, radar antennas emit pulses of radio waves — about 1000 pulses per second, each lasting about a millionth of a second. The waves travel at the speed of light, 300,000 kilometres (186,000 miles) per second, until they strike some reflecting surface, which may be almost anything from solid rock to the water vapour in clouds. The reflected waves are received by the same antenna, in the intervals between the pulses. The time interval between outgoing and reflected pulses is continually translated into visual data, usually numbers on a dial or 'blips' — dots of light — on the screen of a cathode ray tube similar to a TV picture tube.

Radar has a large variety of applications involving precise measurements of distances. It is used for determining altitudes of aeroplanes, navigating in fog and in the dark, and even mapping the cloud-shrouded surface of Venus. A useful, if unwelcome, application of radar is for police speed traps. Here, a special radar device is used, which responds differently to the reflections from moving objects and stationary objects. The greater the speed, the greater the difference.

See also DOPPLER EFFECT; SONAR.

RADAR SPEED CHECKS   *See* DOPPLER EFFECT

RADIANT ENERGY

The form of energy transmitted by electromagnetic waves. A goodly portion of your life is wrapped up in these waves. You see by them (light waves), you are kept warm by them (infrared waves), you are kept informed by them (radio and TV waves), you are medically diagnosed by them (X-ray waves), you are suntanned by them (ultraviolet waves), and, fortunately, you are constantly shielded by the atmosphere from their most lethal form (gamma rays).

The forms and effects may be different, but all electromagnetic radiation shares two aspects: a particle aspect and a wave aspect. One might be tempted to think of radiation as a fast motorboat (the particle) accompanied by the waves it sets up in the water. However, succumbing to such temptation would lead one into error — and did, among early physicists — because there is no water, no medium shaken up by the motorboats, the particles, which are called *photons*.

All electromagnetic radiation, then, consists of particles (photons) accompanied by waves (of what? — of 'waveness' — sorry). Photons differ in the amount of energy they possess. Waves differ in their dimension (wavelength) and in their frequency (number of waves per second). You make use of one of these, frequency, when you tune in a station on a RADIO. High-energy photons are accompanied by high-frequency, short-wavelength waves. Lower-energy photons are accompanied by lower frequency and longer wavelength. Here is a chart of the various groups of electromagnetic radiation, arranged according to their frequencies. The entire chart is called the *electromagnetic spectrum*. Within it is the supremely important narrow band called the *visible spectrum*. This contains the range of frequencies that our eyes respond to, sending messages via the optic

ELECTROMAGNETIC SPECTRUM

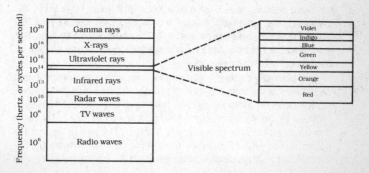

nerves to the brain. Green plants respond in another way to some parts of the visible spectrum, using the light as a source of energy for the food-making process of photosynthesis.

The members of the total electromagnetic spectrum can produce a wide variety of different *effects* – you can see with light waves, you can send communications via radio waves, you can penetrate body tissue with X-rays – but the waves themselves are different only in frequency and wavelength. There is no colour in the light waves that carry a message of red stripes in the American flag. There is no sound in the radio waves that carry a message of music to your radio receiver. It's all done by the receptors (the eye in one case and the radio in the other), built to respond to their particular frequencies. The receptors in the eyes of bees enable them to see ultraviolet in flower petals, a colour that we humans can never see and never imagine. Some lucky (unlucky?) people, when near a powerful radio station, are able to hear its programmes through their tooth fillings! (That's a good point of departure for a science fiction story.)

RADIATION See RADIANT ENERGY

RADIATION (IN CANCER) See CANCER

RADIO

A direct definition of the term *radio* at this point would be too simplistic, too technical, or too long-winded to hold your attention. Let us instead attempt a flanking strategy: turning on your desk or table lamp.

As soon as you turn the switch, electric current begins to flow through the wires in the lamp. At the same time, the wires become surrounded by an invisible electromagnetic field. As evidence of such a field, you may recall a simple experiment (see diagram). Tap-tap the end of a wire against a battery terminal. Watch the compass needle, which is a magnet. It jiggles, driven by the changing magnetic field around the wire. Touch the wire to the battery terminal, and the magnetic field springs up. Take away the wire, and the field collapses. Each up-and-down causes a jiggle of the magnetic compass needle.

You can detect the jiggling effect with the compass several centimetres away from the wire. If you use a more sensitive detector, such as a radio receiver, you will hear crackling sounds. These are induced by the electromagnetic field from the battery wire. The radio can be several metres from the wire. And still farther? To the ends of space; the only limit is the sensitivity of the detection apparatus. For example, the ultrasensitive receivers used for communicating with

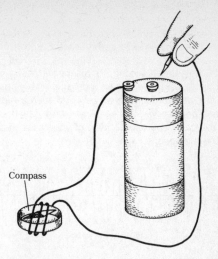

NASA's space probes detect and record radio signals from the *Pioneer 10*, which left our solar system in June 1983. At that time it was 4500 million kilometres (2800 million miles) from the earth. Those signals were emitted by an electric current of about the same strength as the current in the tap-tap compass-needle experiment!

We have crept up on a partial definition of radio: it has something to do with electromagnetic fields. But the tap-tap experiment isn't truly an example of radio because of its random nature – the tap-tap an the resulting uninformative jiggles or crackles.

Formal (not random) radio transmission and reception involve *alternating current*, often referred to as *AC*. This is electric current that flows in one direction, stops, flows in the opposite direction, stops, then reverses its flow again, over and over. Each back-and-forth is one *cycle*; the number of cycles occurring in one second is the *frequency*, measured in *hertz* (abbreviated to *Hz*). Domestic current in Britain is designated as 50-Hz AC, meaning that it does its back-and-forth 50 times per second, (another term is *cycles per second*). In the United States, AC is usually generated at 60 Hz.

Radio and television stations broadcast by sending AC into their aerials. This AC is produced at a variety of rates. For example, Capital Radio in London sends an alternating current of 1,548,000 Hz. This is also known as 1548 kilohertz, or 1548 kHz. That is the number you can turn to on the dial for Capital if you are in or near London. Space on the airwaves is very tight so a number of stations may broadcast at the same frequency. However, the stations are sufficiently distant so that under normal conditions they do not interfere with one another.

Let's observe what happens at Capital Radio as it begins the broadcasting day. The station engineer presses a switch. Immediately an AC of 1548 kHz flows through the aerial, and an electromagnetic field races out into space, building up and falling back 1,548,000 times each second. As diagram A shows, it's a steady, even train of waves. If you turn on your radio at this time, you'll hear nothing, because you can't hear waves of such a high frequency. These *radio-frequency (RF)* waves are also called *carrier waves* because they will carry another type of wave as soon as a sound is made at the microphone.

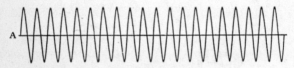

RF (carrier) waves

Now the announcer says 'Good morning'. The sound waves of the voice could be represented as in B. The microphone converts the sound to electric currents. *Audio-frequency (AF)* waves are set up. Their frequency is much less than the carrier wave frequency.

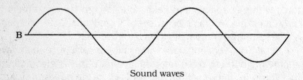

Sound waves

The radio transmitter combines the RF and AF waves into a single new set (C). This new set contains the '1,548,000 cycleness' *and* the 'voice vibrationness'. The combined set is sent up the aerial and out into space in all directions. Some of the waves, travelling at the speed of light, strike your radio receiver. They can enter the circuits

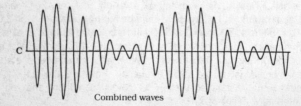

Combined waves

in the radio only if the dial is set at 1548 kHz on the medium wave band. This adjusts an electronic part called a *variable capacitor* to respond to a frequency of 1548 kilohertz. The circuitry in the radio changes the waves to very weak electric currents. These currents are strengthened by a device called an AMPLIFIER and fed into a LOUDSPEAKER.

In diagram A the carrier wave train is steady and even. Then it is changed, or *modulated*, by the shape of the sound waves (B). The change is an alteration of the height, or *amplitude*, of some of the waves, so this method is called *amplitude modulation* — AM for short.

The term *AM* immediately draws your attention to another term, *FM*, which stands for *frequency modulation*. FM is a far superior method of radio broadcasting, because it rejects the crackling static that plagues AM when there is lightning, a passing motor vehicle, or a defective electrical device nearby. Also, FM is better able to carry the subtle curlicues and overtones that we call HIGH FIDELITY. How this is achieved is beyond the scope of this book, but we can afford a few sentences on what FM looks like.

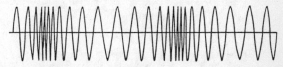

FM waves

Observe that the FM amplitude is constant, while its frequency changes three times. The FM receiver senses the frequency changes and converts them into sound vibrations. Three changes in frequency give three sound vibrations.

Even though the hi-fi aspects of FM are beyond our scope, let's take a brief look nevertheless. Observe the combined AM waves (diagram C). The changes in height represent changes in strength. The tall waves are stronger than the short ones. Under marginal conditions (a weak or distant station, a weak receiver) the shorter (weaker) waves don't get through too well, and the quality of the sound suffers. The FM waves, by contrast, are all of equal strength.

See also FIELD THEORY; RESONANCE; SEMICONDUCTORS.

RADIOACTIVE DECAY   *See* RADIOACTIVITY

RADIOACTIVE DECAY AS ENERGY SOURCE   *See* ENERGY RESOURCES

RADIOACTIVE WASTE See NUCLEAR ENERGY

RADIOACTIVITY

The spontaneous (not artificially induced) breakup, or decay, of certain kinds of heavy atomic nuclei, accompanied by the emission of certain kinds of particles and energy. As defined here, this sounds rather dull and minor, but actually radioactivity is a vivid and highly potent process that began at the first moment of the big bang (see COSMOLOGY), has continued full blast ever since, and should, in some of its aspects, be avoided like the plague. Towards that end, hospitals and laboratories, in areas where radioactivity is at a dangerous level, display the logo shown here.

As to the aforementioned 'certain kinds of heavy atomic nuclei': uranium, protactinium, and thorium are three naturally occurring heavy elements. Their atomic nuclei are unstable; they tend to break apart comparatively easily.

'Certain kinds of particles and energy': as the nuclei decay, they emit ALPHA PARTICLES (helium nuclei), BETA PARTICLES (electrons emitted in fast streams called beta rays), and GAMMA RAYS (short X-rays).

As an example of the radioactive process, consider the first three steps in the decay of uranium.

1. Uranium (U-238) emits alpha particles as it decays to thorium (Th-234) during a half-life of 4.5 thousand million years. The term *half-life* is the time period, roughly equivalent to average time, during which half of the atomic nuclei have completed that step in the process.

2. Thorium emits beta rays as it decays to protactinium (Pa-234) during a half-life of 24.1 days.
3. Protactinium emits beta rays during a half-life of 70 seconds as it decays to another element, which decays to another, and so on, for more than a dozen steps. The half-lives range from millions of years down to thousandths of a second, and the sequence ends with stable, common, end-of-the-line, unromantic lead.

The effects of radioactivity, too, are wide-ranging: from the lingering death sentences emitted by radioactive wastes over hundreds of thousands of years to the destruction of cancerous tissue by radioactive elements in carefully controlled doses of several minutes.

RADIO ASTRONOMY

The study of galaxies, stars, and other celestial bodies by means of the radio waves they emit. This definition can be intriguing and misleading, for it has nothing to do with radio broadcasting or transgalactic telephone conversations. Celestial radio waves, have wavelengths in the same range as short wave, FM, and radar waves, but they carry no coded information (called intelligence). They are similar to the infrared (heat) waves given off by an electric burner set to low. If you could tune in on them, you would hear unintelligible hums, punctuated with squeals, sputters, grunts, squawks, growls, and whistles.

Then what use are they? They are a source of much information to astronomers, increasingly so as the technology for receiving and interpreting them improves. All objects in the universe (except black holes) emit RADIANT ENERGY at various wavelengths; with appropriate instruments, all can be detected, identified, and analysed.

For example, hold your palm close to your forehead, without touching it. Feel the warmth as infrared waves pass between palm and forehead. Than rub your hands briskly for half a minute and try again. Your palms now emit more energetic infrared waves. which your skin feels as a higher temperature. A sensitive receiving apparatus can measure these before-and-after temperatures and even show, on a cathode ray tube, the shape of your face and hands. Special telescopes for night use show warm objects — birds, mammals, car exhausts — as contrasting areas against cooler backgrounds — trees, hills, buildings.

Old-time pictures of astronomers at work show people (always male) peering through glass-lensed telescopes. Human eyes perceive optical waves, but the optical part of the spectrum occupies only a very small part of the total spectrum of radiant energy. So the development of radio astronomy has expanded the field enormously. No longer is the astronomer totally dependent on cloudless nights, almost totally dark surroundings, or clear line-of-sight views. A rough parallel: in the olden days of radio, the aerial had to be out in the open for decent reception, whereas nowadays it can be tucked inside a tiny box inside your pocket, inside an office building.

Here is a brief list of some achievements of radio astronomy:

'Invisible' stars and galaxies Turn off the switch of your desk or table lamp. The white-hot filament of the light bulb ceases to emit *optical* waves, but if you hold your hand close to the bulb, you feel the heat of infrared waves still shining out (or is the verb *shining* appropriate?). As a celestial object cools, it emits radiation in the longer and longer wave brackets. But it continues to emit! Even though our eyes are unable to see infrared waves, radio waves, and many others, the waves are there and, with the proper apparatus, can be observed. (For example, the retinas in the eyes of bees can perceive ultraviolet waves.) Millions of stars are being discovered, one after another, by the use of radio telescopes.

A new look at the solar system The sun is, without a doubt, hot stuff. But there are different degrees of hotness, and these show up as differences in wavelength, easily perceived through radio telescopes. They provide astronomers with much new information about the chemical and physical behaviour of, for example, various regions of the sun. This provides the opportunity to predict changes, such as solar flares and sunspots, that cause magnetic compasses to lie to Boy Scouts and disturb broadcasts with crackling static and distorted pictures.

To the ends of time and space Suppose you had a special talent for seeing wood, anywhere, anytime, out in the open or concealed by other substances. You could look at a house and see the rafters and studs and posts; you could get a reasonably good picture of the structure of the house just by viewing this particular substance, wood. The structure of the universe is now becoming visible by reason of a particular substance: hydrogen. In the 'empty' space of the universe, the most abundant atoms by far are those of hydrogen, each one a tiny emitter of radio waves having a wavelength of 21 centimetres (about $8\frac{1}{4}$ inches). How convenient! Set your radio telescope to receive 21-cm waves, as you might turn you home radio dial to your favourite station. Sweep your aerial back and forth across the sky and you have a picture of the universe's 'empty' space (filled with hydrogen and other sources of radio waves), as well as different spaces filled with 'things', such as stars and planets.

A world full of radio You probably had not realized (and why should

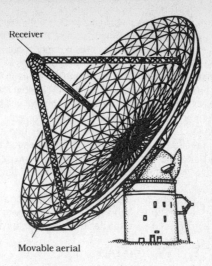

Receiver

Movable aerial

you?) that a tiny atom of hydrogen could be an astronomical object, but so it seems, by any commonsense definition of astronomy. And so is anything else 'out there'. And we can find it by setting our radio telescope's tuner for that particular wavelength!

See also ANALYSIS; STELLAR EVOLUTION.

RADIOCARBON DATING   See UNCERTAINTY PRINCIPLE

RADIO FREQUENCY   See RADIO

RADIO TELESCOPE   See ANALYSIS

RADON

A gas emitted by the natural decay RADIOACTIVITY of radium and uranium present in certain metamorphic rock layers and in the overlying soil formed from such rocks. Radon is considered by many countries to be the second most important cause of lung cancer after smoking. An estimated 1500 people die each year in Britain from lung cancer caused by the gas.

Ordinarily the gas is dissipated by natural movements of the air, but when it seeps from below into houses it tends to concentrate, being 7½ times as heavy as air. In recent years surveys have confirmed that radon is far more widespread in houses than previously thought. A national survey in Britain in the early 1980s pinpointed high-risk areas in Devon and Cornwall; now, many parts of England and Wales are considered at risk. By late 1988 there was no consensus regarding estimates of the number of homes in danger, or as to what constitutes a dangerous level of the gas. One estimate suggests that between 50,000 and 80,000 homes in England and Wales may have radon concentrations over 400 becquerels per cubic metre of air. Continual exposure to this amount gives a radiation dosage of four times that set for nuclear power workers.

In the United States, the Environmental Protection Agency (EPA) has found radon levels in homes of up to 7500 becquerels per cubic metre, 50 times the American 'action level'. Towards the end of 1988 the US government suggested that all buildings should be tested.

It is not known whether radon emission is greater during certain seasons and temperatures, but there is sufficient evidence that the whole problem of radon emission deserves intense study.

RADWASTE *See* NUCLEAR ENERGY

RAM AND ROM

There are two types of computer memory, RAM (*r*andom-*a*ccess *m*emory) and ROM (*r*ead-*o*nly *m*emory). Both are groups of electronic circuits, formed into a CHIP, that store information in a computer. How much information they can store depends on their capacity as measured in kilobytes (1024 bytes), abbreviated *K*. Thus a 64K memory can store 65,536 bytes, each byte being one character (letter, digit, space, etc.). This sentence consists of 35 bytes.

Random-access memory   RAM might be called a computer's internal electronic blackboard. It can be 'written' on, 'read' from, altered, and erased. In fact, if you intentionally or accidentally turn off the electric current to a computer, the information stored in its RAM disappears completely. To prevent such a loss, you can permanently store the data (unless deliberately erased) on magnetic tape, using a tape drive, or on a magnetic disk, using a DISK DRIVE. Any item in the computer's RAM can be looked up in any order, as a book, without having to search through from beginning to end (hence *random* access).

Read-only memory   ROM is the permanent part of a computer's memory. The information stored there can be 'read' but not changed or erased. ROM contains the program that the manufacturer has built into the computer; therefore, it does not need an electric current to keep its programs stored.

Here are some examples of RAM and ROM at work:

- Dividing any number by any other: ROM
- Storing the answer to   74.3   RAM
  107.9
- Deriving the square root of any number: ROM
- Determining the sine of any angle: ROM
- Holding on to the sum of seven numbers: RAM

RANDOM-ACCESS MEMORY   See RAM AND ROM

READ-ONLY MEMORY   See RAM AND ROM

READ/WRITE HEAD   See DISK DRIVE

RECOMBINANT DNA TECHNOLOGY
See GENETIC ENGINEERING

RECTIFIER   See GENERATORS AND MOTORS, ELECTRIC; SEMICONDUCTORS

RECYCLING   See BIODEGRADABILITY

RED GIANT STAR   See STELLAR EVOLUTION

RED SHIFT   See DOPPLER EFFECT

RED SUPERGIANT STAR   See STELLAR EVOLUTION

REFRIGERANT   See REFRIGERATION; OZONE LAYER

REFRIGERATION

A kitchen refrigerator, a domestic air conditioner, and the cooling system in the backpack of a moon-walking astronaut all work on the same principle, cooling by evaporation. Here's how:

If you dip your finger in gin you won't feel anything unusual in the way of warmth or cool. Remove the finger and suddenly it feels cooler. You exposed your wet finger to air, and the gin evaporated — changed to a vapour. A change from liquid to vapour requires heat, which the alcohol took from your finger. Warm finger loses heat, becomes cool finger.

But you've also lost the gin, into the air. If you catch the vapour and change it back to a liquid, to use over and over again, you'd have a refrigerator working on the evaporative cooling cycle. Most coolers and freezers work on such a cycle, like this:

1. A special, easily evaporating liquid, the refrigerant (usually *Freon*, a mixture of carbon, hydrogen, fluorine, and often chlorine) is pumped into a coiled pipe called the *evaporator*.
2. The liquid can't escape into the air, because it's inside a pipe. But it can evaporate.
3. The evaporation of the liquid refrigerant into vapour causes the coiled pipe to become cool. The cool pipe in turn cools the surrounding air, water, milk, or whatever.
4. The refrigerant vapour then flows back through a pipe into a pump, the pump squeezes it (compresses it) into a hot vapour that then becomes a hot liquid. A fan blows air across the tube and cools the liquid. Now the cool liquid refrigerant is ready for its next round trip through the machinery.

REGULATORS (DNA)   See DNA AND RNA

REJECTION OF TISSUE   See IMMUNE SYSTEM

RELATIVE HUMIDITY

A measure of the moistness of air at any particular temperature, compared with the maximum it could hold at that temperature. Warm air can hold more moisture than cold air. Thus, if a certain package of cold air (say 2°C) is holding all the moisture it can possibly sustain, we say it is 100% humid. If that package is warmed, its capacity to hold moisture is increased; it has become less than saturated, less than 100% humid. In warm weather we feel more comfortable when the relative humidity is low because perspiration evaporates more readily then, a cooling effect.

RELATIVITY

The relativity theories of Albert Einstein (1879–1955) deal with the most fundamental descriptions of the physical universe: the concepts of time, space, motion, mass, and gravitation. To attempt to explain (rather than describe) them here would be an act of presumption, for two reasons: (1) Much heavy mathematics is involved, and (2) the concepts of relativity are not readily accessible with our (relatively) crude experience, where the weight of a grain of salt is considered small, where a rocket's escape velocity — 11.2 kilometres (7 miles) per second — is regarded as very high, and where a clock gaining a second in ten years is deemed superaccurate.

RELATIVITY

The ideas of relativity emerge only at the boundaries of such a world, in the domains of the supersmall, the superfast, the superlarge, the supermassive. Yet the proofs of relativity's seemingly wild assertions are quite evident and useful to physicists working with particle accelerators, astronomers studying galaxies, and mathematicians calculating the orbits of spaceships.

First, to help us cast loose from prerelativity thinking, we will engage in a 'thought experiment'. Theoretical physicists delight in these.

Resting or moving? You're sitting in a stationary train, reading. Alongside is another train, also stationary. After a while you look up and notice that the other train is in motion — or is it your train, gliding silently — or are both trains moving at different speeds? Use your camera to take a time exposure of the other train. You'll get a streaky picture — but will it tell you about the *absolute* motion of either train or only the *relative* motion between them?

There is no absolute 'stationariness' (rest state) or absolute motion. If you choose to label yourself as the single rest-state object in the entire universe, with everything else revolving about you, so be it (confirming your doting grandparents' insistence that the sun rises and sets on you). Think of it this way: (1) Suppose there are only two objects in the entire universe, and (2) the distance between them is increasing steadily in a straight line. Then (3) is it possible to say which one (or both) is moving? How about three objects? Three million?

There is no absolute rest, no absolute motion, *with one exception*, which shall be revealed at its logical place in this discourse.

Space: absolute or relative? Onwards in our quest for absoluteness, seeking an unvarying, 'unrelative' place or standard of measurement. Here's a thought experiment similar (but without heavy mathematics) to one proposed by Einstein:

Part 1: A train is waiting at a station platform. A passenger lays a metre rule against a window, fore and aft, and finds that it fits exactly; the window's width is 1 metre. Outside, an observer on the platform measures the same window. Both measurers are stationary in relation to each other; physicists say they are in the same *frame of reference*. Both get the same result: 1 metre.

Part 2: What if the measurements are taken in different frames of reference, one stationary and one moving? Back up the train and send it forwards. As it passes the station, the window is measured again, by the passenger from the inside and by the observer from the outside (it can be done in a split second but needs paragraphs to describe and explain). The result from the inside is no surprise. 1

metre, because window, metre rule and passenger are moving together. They are in the same frame of reference, as when all were standing still.

But from the platform outside, the window measures less than 1 metre! Only a tiny bit less at a 95-kilometre (60-mile)-per-hour-crawl, but at 262,318 kilometres (163,000 miles) per second ($7/8$ the speed of light), the distance from front edge to rear edge is only half as long as during the rest measurement. Is the passenger aware of it? No, because his metre rule still fits neatly in the space between the window edges.

So distance, which is constant *within* the same frame of reference, becomes relative *across* two different frames of reference. And it works both ways! If the passenger in the high-speed train measures the window of a stationary train on the track alongside, it too measures less than a metre wide.

Bewildering but true. Although (at present) there are no trains whizzing along at such speeds, there are orbiting artificial and natural satellites, planets, and vast systems of stars within galaxies, all travelling at high speeds. They all bear out Einstein's statement that distance — the measurement of space — is relative.

Time: absolute or relative? 'Dr Bagley can see you on Thursday at 2:30'. Dr Bagley's secretary is making a four-dimensional appointment for you. The office is located at the corner of East Street and North Avenue (two dimensions) on the second floor (third dimension) at 2:30 (fourth dimension). Barring accidents, the appointment will be kept, because all four dimensions are stable. The office will remain fixed in its three dimensions of space, and Dr Bagley's watch and your watch will continue to count time at the same rate. All this pleasant reliability is the result of inhabiting the same frame of reference as Dr Bagley.

What if you inhabited a different frame? In the example of the train whizzing by the railway platform, you learned (or, more likely, took on faith, hesitantly) that a metre rule shrinks fore and aft when measured by an observer in a different frame of reference. Not noticeably at customary speeds, but significantly at astronomical and atom-smashing velocities. The three dimensions of space are constant only when measured within the same frame of reference.

Now consider the fourth dimension, time. Is it also relative to frames of reference? Should a pair of perfect clocks, showing exactly the same time, continue that precise way if one clock is taken for a ride on a space shuttle at 30,000 kilometres (18,000 miles) per hour while the other stays at home? Experiments of this kind have been done, and a tiny but exact difference was measured. Clocks in motion — any kind of clocks, in any kind of motion — run more slowly than clocks at rest. There is no absolute interval of time. *Time is relative.* Something about motion causes the measurement of time to slow down.

Here's an example of time's relativity. Outer space swarms with cosmic rays, which are mainly extremely energetic protons. Some, it is believed, are fired out of the sun, others out of exploding stars called supernovas. When cosmic rays strike our upper atmosphere, they shatter some nitrogen molecules in the air. A series of spontaneous split-ups ensues. In one part of the series, muons are converted into electrons. This change has been timed under laboratory conditions with stationary apparatus; it takes about 2 millionths of a second. But at high speeds, streaking down from the upper atmosphere to the ground, the change is much slower, so that many muons actually reach the ground intact, a moment before the change. Motion and time are inversely related: the faster the motion, the slower the passage of time.

Mass: absolute or relative? Another surprise introduced by relativity theory is the concept of *relative mass.* Mass is the property we earthlings associate with weight; mass is proportional to the amount of 'stuff' — protons, neutrons, electrons, and other particles — in an object. Yet mass is not constant. Mass varies with the amount of energy in it. Here's a down-to-earth example:

On the earth there lies a football, unmoving. To change it into a moving object, we must put energy into it (e.g., kick it). When this happens, the mass of the football is increased! Again, by only a tiny percentage when a football is kicked, but by much more when an electron is fired into motion at several thousand kilometres per second. This happens, for example, when a TV switch is turned on. In a TV picture tube, the electrons gain about 2% in mass. And in a high-powered particle accelerator, electrons gain as much as 40,000 times their rest mass.

Add energy to an object (e.g., kick a football, wind a windup clock, heat a frying pan), and it gains mass; subtract energy, and it loses mass. We could venture to say that mass and energy seem to be equivalent — interchangeable, in a manner of speaking. In fact, Albert Einstein did venture to say exactly that: $E = mc^2$. E stands for energy, m stands for mass, and c^2 stands for the square of the velocity of light. To put it in less technical terms, if an aeroplane ticket (a mass of paper, weighing perhaps 15 grams or ½ ounce) could be totally converted into energy, it would be enough to propel an aeroplane several thousand times around the world.

So mass is not a fixed property of an object; it depends on (is relative to) what's happening to it. *Mass is relative*.

Gravitation: absolute or relative? Here is another aspect of the strange world of relativity (which is simply your familiar world of motion, space, time, and mass viewed at its extreme edges). Gravitation is a force that can be described according to (relative to) an observer's viewpoint. We can check that statement in a lift with a set of bathroom scales. Stand on the scales and they show your weight. Press the 'up' button; as the floor begins to rise, it presses upwards against the scales (and against you) and you seem to be heavier. If there were no windows in the lift (common enough) and if you had just been placed on the scales whilst asleep (uncommon), would you have any way of knowing, upon awakening, whether you were inside a very silent lift beginning to rise or in a stationary box on a planet more massive than the earth? Is it gravitation that's showing on the scales or inertia — the reluctance of a body to being budged, speeded, or slowed (the word physicists use is *accelerated*) — or some of each? Stated more elegantly, gravitational mass and inertial mass are indistinguishable — or perhaps two aspects of the same property. So there's no fixed 'unrelativistic' way to describe gravitation; it's relative. And so, too, are the previously mentioned aspects of the physical universe: motion, space, time, and mass.

Velocity of light: absolute or relative? Three travellers are comparing notes. Traveller A tells about driving past a busy archery range. He comments on how swiftly the arrows flew. Traveller B says, 'I too, was there, and the arrows were so slow they almost seemed to be standing still.' Traveller C says, 'I saw them travelling at normal arrow speed.'

Traveller A was driving against the direction of the arrows' flight. Traveller B was driving in the same direction of the arrows' flight. Traveller C had stopped his car before observing the arrows. He had put himself in the same frame of reference as the archery range; the others had not.

The speed of a moving object depends on (is relative to) whether it is measured in the same frame of reference or across two frames. This is true for measuring the speed of arrows, bullets, and snails — *everything but particles of light*, called photons. No matter how fast or slow an observer moves, or in what direction, his measurement of the speed of light comes out the same, 299,792.8 kilometres (186,282.5 miles) per second. It seems impossible, but there it is. The speed of light is *not* relative, but *absolute*, independent, not related to the motion of the observer or the source, and independent of the common sense by which most of us operate.

RENEWABLE ENERGY SOURCE

An energy source that is not depleted by continued use. Winds turning windmills are an example.
See also ENERGY RESOURCES.

REPLICATION OF DNA   See DNA AND RNA

RESOLVING POWER

A measure of the *sharpness* of a scientific instrument. You are most likely to encounter the term in reference to optical instruments. A telescope, microscope, or camera lens of high resolving power is more expensive than a lens of low resolving power because it it more difficult to design, manufacture, and test.

Resolving power, or resolution, is different from magnifying power, which merely tells *how much*, not how well, an instrument magnifies. With a telescope having high magnifying power but low resolving power you can see the moon's craters, but the details will be coarse, as if drawn with a soft crayon rather than with a sharp-pointed drawing pencil.

Resolving power is described mathematically in terms of seconds of arc (one such second is $1/3600$ of one degree). In a simpler form, it can be stated as the maximum distance at which a lens can show two close-together parallel lines as two *separate* lines rather than as one fuzzy one. For example, a very high resolution telescope lens 305 millimetres (12 inches) in diameter can show these separate lines, ||, at a distance of 800 metres (½ mile). At that distance, with such a telescope, you could read this book.

RESONANCE

Vibrations set up in one object, caused by vibrations received from another object, such as the chattering of a loose bumper on a car, resulting from the vibrations of the engine. Resonance is usually thought of in connection with sound energy, but it applies in any case where energy is expended in a regular, repeating manner — for example, in radio waves, the movements of a playground swing, or the vibrations of atoms and molecules.

A simple demonstration helps to explain the resonance principle. Four sugar cubes are suspended from threads A, B, C, and D. These in turn hang from a single thread. Start A swinging, and C (same length as A) will begin to swing in resonance with it. B and D may move slightly, but only in a fitful, uncertain way. If you adjust the lengths of strings B and D, you can get them to resonate too. All

RESONANCE

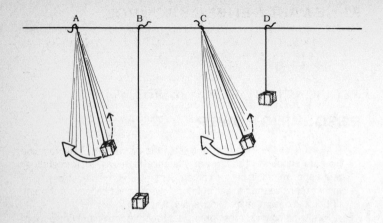

things that can move in a regular, repeating way have a natural 'swing' or period of vibration.

The resonance principle has many uses:

As a method for selecting a period of vibration Every radio station or channel is assigned a particular frequency for the waves it broadcasts. For example, Radio 4 uses radio waves with a frequency of 198,000 hertz (cycles per second). When you turn the dial of your receiver to 198, you are adjusting an electronic part, called a *variable capacitor*, to resonate at 198,000 hertz. It's like changing the length of string D (your radio receiver) until it resonates to string A (the radio station).

As a means of transferring energy When you give someone a ride on a playground swing, you can reach monumental heights by timing your pushes to the natural frequency (swing rate) of the swing. You are making yourself into a vibrator (string A) and transferring energy into the swing and its passenger (string C). With sufficient enthusiasm and well-timed energy, you can achieve resonance to the point of near-disaster.

This is the principle of ULTRASONICS, as applied to dentistry, jewellery cleaning, and many more exotic tasks. Normal audible sounds are in the range of 20 to 20,000 hertz (Hz). Beyond that comes ultrasound, the frequency range of many tiny objects. For example, molecules of grease and dirt can be made to resonate to a frequency of about 25,000 Hz by energy transferred from an ultrasonic generator. The molecules vibrate with increasing violence until they are shaken loose. The same principle applies to an ultrasonic dental instrument used for removing plaque from teeth. A similar method, using radio waves, is employed in diathermy machines to generate heat in body tissues and to cook food in microwave ovens.

As a method of analysing substances The nucleus of an atom is surrounded by electrons revolving in orbit. The movements of the electrons and corresponding movements of the nucleus set up a vibration that is different for each different kind of atom. Oxygen atoms, for example, have a different vibrational frequency from nitrogen atoms. So one way of identifying substances, even in very tiny amounts, is by measuring this frequency. The process requires the use of magnetic fields (see *nuclear magnetic resonance (NMR)* at ANALYSIS; MAGNETIC RESONANCE IMAGING).

As a much simplified parallel to the NMR technique, consider the following arrangement. There's a pendulum inside a box, swinging constantly. You are asked to determine the pendulum's frequency without opening the box. Fortunately, there is a second pendulum suspended from a string whose length you can change. Set the pendulum swinging. Does it jiggle a bit, or does it develop a full, healthy swing — that is, does is resonate to the inside pendulum? If it just jiggles, try another length, and another. When you achieve resonance, count the frequency of the outside pendulum, and you'll know the frequency of the one in the box!

The pendulum swinging inside the box is like the vibrating atom. The outside pendulum, whose frequency is controllable, is like the electronic *oscillator* in the NMR machine. The oscillator generates radio waves whose frequency can be regulated, and it beams them at

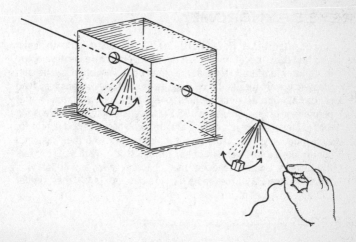

the substance to be identified. If the radio waves are not in resonance with the vibrating atoms, a weak jiggly line is seen on the TV-like screen of an *oscilloscope*. If the frequencies are in resonance, a strong peak is seen. The peak shows the frequency of the atoms in the sample. Because a particular frequency is unique for a particular substance, the identity of the sample is thus known. Many other analytic methods are available, but one special advantage of NMR is that it can be used on living things, from single cells to whole bodies, without harming them. It can even follow the progress of a particular substance through a series of chemical changes within the living organism.

RESTRICTION ENZYME   See GENETIC ENGINEERING

RETRIEVAL

Withdrawing data from the memory of a COMPUTER (from Old French, finding again). Information is stored in various kinds of memories, chiefly electronic and magnetic. A computer's retrieval rate — how fast it can search through its memory and come up with the desired item — is an important factor in its speed of operation. Thus if the information is stored on tape, the retrieval rate is necessarily slow, because the tape must be scanned serially until the desired item is found (like finding a particular bar of music on a cassette tape). Retrieval from a disk is much faster, like moving the tone arm of a record player to a selected groove on a record. Still faster is retrieval from the electronic memory inside a computer, where nothing moves except electric current, at many thousands of kilometres per second.

REYE'S SYNDROME

Named after R.D.K. Reye (1912–1978), the Australian pathologist who first described it in 1963, a rare, dangerous disease developing in children suffering from a viral infection. The brain swells and presses against the skull. Vomiting, fever, and convulsions may lead to coma. One in four victims dies, and survivors may suffer permanent brain damage. In 1981 researchers noted an increase in the incidence of Reye's among children who had received asprin for viral infections such as flu or chicken pox. Further research confirmed the involvement of asprin; one study, by the US Centers for Disease Control, showed that children given asprin for viral infections were 25 times more likely to develop Reye's than similar children not given asprin.

RHEOSTAT   See GENERATORS AND MOTORS, ELECTRIC

RHEUMATOID ARTHRITIS See ARTHRITIS

RHYOLITE See ROCK CLASSIFICATION

RIBONUCLEIC ACID See DNA AND RNA

RIBOSE See DNA AND RNA

RIBOSOME See DNA AND RNA

RICHTER SCALE

A calculation of the intensity of an earthquake at its source based on instrumental records. Also called Richter magnitude, it is named after Charles F. Richter, the American seismologist who devised it in 1935. The scale is logarithmic (to base 10); each step up represents an intensity 10 times as great as the previous step. Thus a magnitude 4 earthquake is 10 times as powerful as one of magnitude 3 and 100 times as powerful as one of magnitude 2 (10 times 10). The Richter scale has limited value as a measure of destructiveness, which depends also on other factors such as the distance of population centres from the earthquake's centre and the type of building construction involved.

The most powerful earthquakes, at a magnitude of nearly 8.8, occurred in Chile in 1906 and in Japan in 1933. The famous San Francisco earthquake of 1906 had a magnitude of 8.3.

RIFT VALLEY See PLATE TECTONICS

RIGHT ASCENSION See CELESTIAL COORDINATES

RNA See DNA AND RNA

ROBOTICS

The study of the design and use of robots (Czech *robota*, compulsory service), machines programmed to carry out a series of operations without human guidance. (The machines didn't ask for the job, hence 'compulsory'.) Robots are favourite characters in science fiction, but you probably possess a few borderline cases at home: heating systems that are programmed according to time and temperature; alarm clocks that stop their buzzing or ringing after you refuse to acknowledge their presence; and video-recording machines that tape one of your favourite programmes while you are watching another of your favourite programmes on TV.

Computer-controlled robots are used in industry to do welding, assembling, and machining and to handle various materials.

ROCK CLASSIFICATION

We are about to compress a term of geology into a few paragraphs, thus incurring the wrath of the gods Vulcan and Thor and of professors of geology.

To a geologist, rock is the material composing the outer part, the *crust*, of the earth. Most rock is hard, but there are exceptions, such as talc, soapstone, clay, and volcanic ash. Rock is generally a mixture of minerals; *granite*, for example, is composed of quartz, feldspar, mica, and hornblende. Some rock is a single mineral: *marble* consists only of calcite (a pure white form of calcium carbonate) with 'impurities' that stain it with various colours and veins.

Rocks can be classified according to their mode of origin.

Igneous rock Igneous (Latin *ignis*, fire) rock is sometimes called *primary rock*. It is the original stuff, molten minerals (*magma*) welling up from the depths (acquiring a new name, *lava*, when at the surface), which then cools and solidifies. Whilst solidifying it forms into crystals, whose size and appearance depend mainly on the rate of cooling. Slow cooling creates large crystals and a very grainy texture. Faster cooling causes smaller crystals and finer grains. Still faster, and the crystals are so tiny that no graininess is evident, just an overall glassiness. Thus, for example, a certain mixture of the minerals feldspar, quartz, mica, and hornblende in a magma can cool into granite (coarse-grained), *rhyolite* (finer-grained), or *obsidian* (glassy). The rate of cooling depends on what happens to the magma. If it forces its way up and out, it cools rapidly and builds up

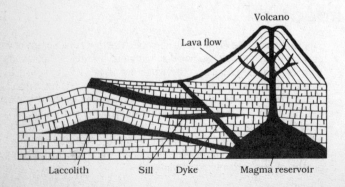

an *extrusive formation* (Latin *extrudere*, thrust out) — either a cone-shaped structure (a *volcano*) or a long cliff or wall. Sometimes the molten rock doesn't quite make it all the way to the surface but spreads horizontally beneath, making *sills, dykes, laccoliths*, or other *intrusive formations*. These, blanketed by rock layers, are likely to cool slowly. Two common types of intrusive igneous rock are granite and gabbro.

No sooner does the extruded igneous stuff harden than the forces of *erosion* (rain, snow, wind, heat and cold, and airborne chemicals) begin to degrade it, in the process called *weathering*.

Sedimentary rock Broken up by erosion, humbled by degradation, the igneous rock turns into boulders, rock fragments, gravel, sand, and silt, but its story isn't over — it never is. Big and little, the rock particles roll down, are carried by glaciers, or are pushed by wind and water. The boulders and rock fragments usually end up helter-skelter near their origins. The lighter material is carried longer distances, but it too is eventually dropped, deposited in layers — sediments (Latin *sedere*, to sit, to settle). Usually the sediments become compacted and hardened into sedimentary rock, due to the action of one or more of three forces: pressure, caused by the weight of more and more sediment settling down; heat, caused by the pressure and by the heating from the magma below; and the action of various chemicals, especially in seawater, that act as binders, like the cement in concrete.

Sedimentary rock can usually be identified by its layered (stratified) construction. Each layer indicates a period of time — years, perhaps or centuries — when conditions were different from those that formed the layer above or below. There is history in the layers of a pebble, like the history in tree rings. Sandstone, shale, and soft (bituminous) coal are examples of common sedimentary rocks. In some places pebbles were cemented in with finer sediments, resulting in rocks with a fruitcake appearance. Breccia is an example.

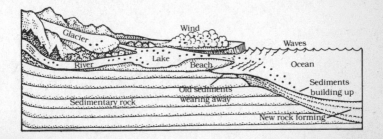

Metamorphic rock This is has-been stuff on the way to becoming something else. For example, limestone, a sedimentary rock mostly formed from the skeletons of tiny sea animals, gradually metamorphoses (Greek, to change form), under the effect of heat and pressure, into marble. It has the same chemical composition, calcium carbonate, but it is denser, harder, more compact. Gneiss, schist, slate, and hard coal (anthracite) are all metamorphic rocks.

Metamorphic rock isn't necessarily the end of the line. Deep in the earth's crust the rock may be subjected to temperatures high enough to melt it, forming magma. The magma, cooling, may form igneous rock, continuing the *rock cycle*.

ROCK, IMPERMEABLE AND PERMEABLE
See GROUNDWATER

ROCK FAULT See FAULT

ROM See RAM AND ROM

ROTOR See GENERATORS AND MOTORS, ELECTRIC

RUDDER

A moveable control surface at the tail of an aeroplane or glider, used to steer the aircraft by moving the tail to the right or left, similar to the rudder on a ship.

See *also* AERODYNAMICS.

RUMBLE

A low-pitched sound emitted by a record player, especially when the volume is turned up high. Rumble is caused by the motion of various mechanical parts, such as the motor, turntable, and belt. There is almost no rumble in a well-designed record player.

See *also* FLUTTER.

SACCHARIN See SWEETENING AGENTS

SARCOMA See CANCER

SATELLITE See SOLAR SYSTEM

SATURATED FATS See CHOLESTEROL

SCANNING

In reading this page, your eyes sweep, or *scan*, from left to right along the first line, gathering information (words or phrases). The eyes flick to the second line, repeating the scanning process, and so on, line after line.

The scanning technique is part of the operation of various electronic devices, such as radar, the CAT scan and the scanning electron microscope. The most familiar example is undoubtedly television. In the TV camera, a thin beam of electrons scans a scene line after line, converting information (light from the scene) into electrical impulses. In a TV set, a similar electron beam, in step with these impulses, sweeps the inner face of the TV picture tube line by line; the pattern of impulses is converted back to dots of light.

SCANNING ELECTRON MICROSCOPE
See MICROSCOPE

SCIENTIFIC LAW See SCIENTIFIC TERMS

SCIENTIFIC TERMS

Scientific terms are often misused, innocently by some, deliberately by others such as purveyors of 'scientifically researched' slimming remedies, psychics, astrologers, and creationists. Here are a few basic terms whose definitions, accepted by scientists, are common to all branches of science.

Scientific law A description of a regularity in nature, observed many times and found to have *no* exception. 'When a spherical object is released at the upper end of an inclined plane, it rolls towards the lower end' is not a law. Although it holds true for tennis balls on earth, it doesn't work for helium-filled balloons or in spaceships. 'The pressure of a confined gas increases with a rise in temperature' *is* a law, stated in mathematical terms as the ideal gas law, a combination of Boyle's law and Charles's law. It has been confirmed by countless laboratory experiments, without exception, and unhappily by increases in the frequency of pneumatic tyre blowouts on hot days.

A scientific law moves forward into *prediction* (scientific, not psychic), enabling application of the 'regularity in nature' to new conditions. Thus measuring the regular movement of the moon and earth enabled Newton to calculate the mass (weight) of the moon relative to the earth's. From this he was able to predict the surface gravity on the moon. This enabled present-day astronomers to estimate the total of load (spacesuit, equipment, etc.) an astronaut would be able to carry comfortably while walking on the moon. The moon's surface gravity is only about one-sixth that of earth.

Experiment An attempt to put a frame around a piece of the physical world — an electron, an ant colony, an aeroplane wing — in order to observe it in detail and (usually) to test the effect of imposing a change (a variable) on it. 'How does a 10% decrease in soil moisture affect the growth of this newly developed breed of cauliflower?' is a valid question for an experiment. So is 'How does a 10% decrease in sunlight affect ...?' But 'How do both together affect ...?' is not, because two variables cannot be tested simultaneously with valid results.

Not all experiments involve total control by the experimenter. For example, observations for studying the chemistry, sizes, temperatures, and other aspects of stars are true experiments, although the astronomer doesn't expect to insert variables into the objects under observation.

Theory An often-misused word, as in 'Evolution is only a theory', with the 'only' implying a shakiness, as if some props were missing. Quite the opposite. A theory is an explanation of a fundamental relationship *that has been supported by experiments*, with no exceptions found. (Remember, the term *experiment* involves systematic observations, as in stellar measurements.) Scientific laws are usually explained by theories. The gas laws are explained by the *molecular theory*, which states that gases are composed of molecules in motion, producing pressure; their motion is directly related to their temperature.

Hypothesis A tentative explanation of a set of facts, waiting for verification by experiments. Fact: in many parts of the world we find fossils of deep-sea fish in layered (sedimentary) rock formations thousands of metres above sea level. Hypothesis: these rock layers were formed at the sea bottom and later lifted, becoming mountains. Experiment: seek evidence that (1) rock layers are laid down at the sea bottom and (2) rock layers can be lifted to become mountaintops.

Technology Scientific knowledge applied to solving real-life problems. (We hasten to add that these problems are of a nature that can be tackled by scientific means.) Consider the problem of achieving nuclear fusion to obtain electrical energy from ordinary water at a hundredth the cost of conventional sources. The scientific theory is clear and has been supported by a number of laboratory experiments, but the technology for doing it on an economically feasible scale is yet to come (see NUCLEAR ENERGY).

Placebo A trick, in experimenting with human beings, to ensure that a psychological variable has not been introduced. The test material (say, a presumed remedy for arthritis) is given to some subjects, while a similar-appearing but neutral substance, the placebo, is given to others. Another use of the placebo (Latin, I will please) gives it its name: an inert medicine used to assure a patient that his or her needs are being attended to when in fact no true medicine is really needed.

Control The 'untreated' group of subjects in an experiment. Take the case of an agricultural scientist who is testing the efficacy of a newly developed synthetic fertilizer for wheat plants. The new fertilizer is applied to an acre of plants (the experimental group), while another acre of plants is treated with an established fertilizer whose efficacy is known (the control group).

Double-blind test A way of doing placebo-type experiments while keeping human bias, conscious or unconscious, from creeping in. The person actually administering the test substance and the placebo has no knowledge of which is which; only the experimenters know.

Model A portion of reality selected and arranged so as to study one or more significant factors within it. For example, an unusual number of hurricanes sweep through a hitherto calm tropical area. Meteorologists compile a mass of relevant data (pressure, temperature, wind force, etc.) taken before, during, and after each hurricane, attempting to discern a common pattern – a model – that may help them to forecast future hurricanes.

SCIENTIFIC TERMS

A model may also be an *analogue*, possessing features in common with a problem being explored. Although high tides and low tides occur throughout the world, there are some places with no measurable tides, while not too far away there are normal tides. Why? We can't do an experiment with the world's oceans, but we can examine one essential aspect of ocean tides: they are a regular, repeating rise and fall — a vibration. Then we find another source of vibration that we *can* handle — a model (Greek, resembling, proportionate). How about a vibrating string? Pluck a tight music string in the middle, and it vibrates as a whole. But pluck it elsewhere, and the string vibrates in sections, *with places that don't vibrate at all!* They are called *nodes*, and the oceans are full of them. The plucker of tides is the gravitational force between the rotating earth and the sun and moon. The breaker-up of whole vibrations is the interference by coastlines and islands.

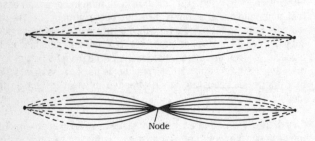
Node

Basic research Scientific investigation performed for the purpose of advancement of knowledge per se, without regard for its immediate or ultimate practical applications. The foregoing sentence may strike you as somewhat on the glib side, and it is. Actually, there's a grey area between basic and applied research, swarming with curious little exceptions, human foibles, and motivations. (The preface of *The Double Helix* by Nobel Prize winner James D. Watson is illuminating reading.)

Deduction The logical process by which a general statement (such as a scientific law) is applied to generate specific information. The ideal gas law is a general statement: it relates pressure and temperature in all gases: When applied to a specific gas under a special set of temperatures, it enables us to deduce a corresponding specific set of pressures. Then we can apply these deductions practically in, say, determining the required wall thickness of a steam boiler.

Induction The logical process whereby a series of related observations and/or calculations are employed to build a general law. It is perhaps the earliest form of intellectual operation:

> Observation: 'When I pushed that stuffed bear off the table, it went down'. Observation: 'When I pushed that ball off the table, it went down'. (And so on for numerous observations.)
> Induction to general law: 'When I push things off the table, they go down'.
> Application of general law (deduction): 'Mother has just placed a bowl of cereal in front of me. I shall push it'.

Scientific research can never be totally inductive or totally deductive, but induction is the logical operation most frequently employed by mature thinkers, with hopes for occasional leaps into *serendipity*, making fortunate and unexpected discoveries by accident. The term was coined in 1754 by Horace Walpole after the characters in the fairy tale *The Three Princes of Serendip*, who made such discoveries.

SCIENTIFIC THEORY See SCIENTIFIC TERMS

SDI See STRATEGIC DEFENSE INITIATIVE

SEAFLOOR SPREADING See PLATE TECTONICS

SECONDHAND SMOKING See SMOKING

SEDIMENTARY ROCK See ROCK CLASSIFICATION

SEISMOLOGY See PHYSICS; RICHTER SCALE

SELECTRON See GRAND UNIFIED THEORY

SEMICONDUCTORS

Materials that conduct electricity moderately well – not as well as the metals (copper, aluminium, iron, etc.) and not as poorly as the insulators, or nonconductors, (rubber, glass, most plastics, etc.). Semiconductors are basic in the operation of almost every electronic device, such as computers, TV, cardiac pacemakers, and radios. There are many kinds of semiconductor materials, but the most common is the crystalline element silicon. In fact, certain regions of California and Scotland where the electronics industry is particularly heavily concentrated have been dubbed Silicon Valley and Silicon Glen respectively.

SEMICONDUCTORS

The supreme value of silicon and similar semiconductors is that they make possible the design and manufacture of very small, very complex, and yet very cheap electronic circuits. A £10 calculator contains several thousand circuits, each with one or more switches and other working parts. A home computer contains a million or so. What goes on in these circuits? How can so many be compressed into such a small space?

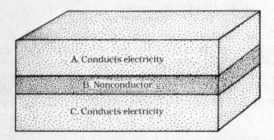

The principal working part of each of these circuits is a *transistor*, which is a kind of silicon sandwich. The 'filling' and the 'bread' are made of silicon that has been doped (altered) by the addition of other substances. Doping alters the conductivity of the slices. One slice (A) welcomes the entrance of electric current from the outside into the filling; the other (C) promotes the exit of current from the filling to the outside. The filling itself (B) is almost nonconductive unless it is receiving a private current of its own from the side. The net result is this: firstly, when the filling gets no private current, no other

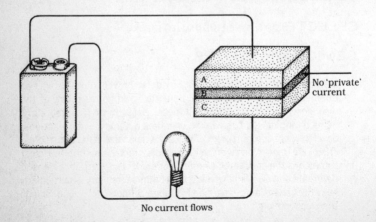

SEMICONDUCTORS

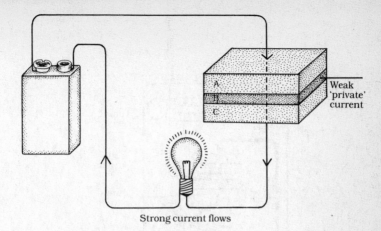

Strong current flows

current can pass through the transistor; secondly, when the filling does receive its private current, it allows other current to flow all the way through the transistor. In effect, this is an electrically operated switch. Because there are no moving parts, only moving currents, the device can be made very small. A chip the size of a postage stamp can hold a million transistors.

Transistors are used for three main purposes:

On/off switches For example, pressing a number key on a calculator or computer sends current to the 'fillings' of transistors that represent that number. Those transistors allow current to flow to a memory section, where other transistors are turned on to hold the number, awaiting your next move, and to the section that displays the number, in the form of bars that are lit up or darkened. In the same way, all the arithmetic operations are performed by the turning on or off of numerous transistor switches.

Rectifiers Most electronic devices are built to plug into alternating current (AC), but most of the circuits inside can work only on direct current (DC). Specially doped semiconductors (diodes) conduct current in only one direction: they *rectify* the current, turning AC into DC.

Amplifiers They convert weak electric signals into strong ones. For example, broadcast radio waves set up weak pulses of electric currents in your radio. The currents must be amplified to operate the radio's loudspeaker. The currents first pass through a tuner, which enables you to select only the current from the station you want,

SENILE DEMENTIA

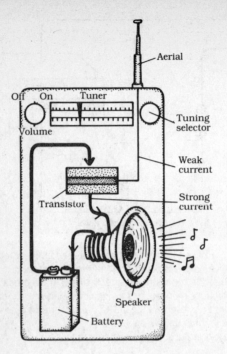

blocking out all the others (see RESONANCE). Then the selected feeble currents (the 'private current' mentioned earlier) flow into the 'filling' of a transistor. There, each pulse causes the transistor to conduct one pulse of strong current from a battery in the radio. The strong pulses operate the loudspeaker.

SENILE DEMENTIA   See ALZHEIMER'S DISEASE

SENILE PSYCHOSIS   See ALZHEIMER'S DISEASE

SENILITY   See ALZHEIMER'S DISEASE

SEPTUM OF THE HEART
See OPEN-HEART SURGERY

SERENDIPITY   See SCIENTIFIC TERMS

SEX HORMONES   See STEROIDS

SHEET LIGHTNING   See THUNDER AND LIGHTNING

SHEET STEEL ROLLING See MANUFACTURING PROCESSES

SHINGLES See HERPESVIRUS DISEASES

SHOCK THERAPY See ELECTROCONVULSIVE THERAPY

'SHOOTING STAR' See SOLAR SYSTEM

SICKLE-CELL ANAEMIA

A hereditary blood disorder.
See also GENETIC DISEASES.

SIDE CHAINS See PROTEINS AND LIFE

SIGMA PARTICLE See ELEMENTARY PARTICLES

SIGNAL-TO-NOISE RATIO See NOISE

SILICON See SEMICONDUCTORS

SILICON DIOXIDE See QUARTZ

SILL (GEOLOGY) See ROCK CLASSIFICATION

SIMPLE MICROSCOPE See MICROSCOPE

SINOATRIAL NODE See CARDIAC PACEMAKER

SMALLPOX See IMMUNE SYSTEM

SMART ROCK See STRATEGIC DEFENSE INITIATIVE

SMOKING

A 1988 report of the surgeon general of the United States affirmed what many researchers had maintained for years, that the nicotine in tobacco is an addictive drug, 'as addictive as heroin and cocaine', in the words of the report. Numerous scientific studies make clear that cigarette smoking triples the danger of sudden cardiac death, and multiplies the risk of emphysema 11 times and of lung cancer up to 25 times. It increases mortality in the fetuses and infants of smoking mothers, and it decreases life expectancy. It is the cause of an estimated 100,000 premature deaths per year in the United Kingdom alone.

These dangers are not confined to smokers. Recent research has established that merely breathing smoke-laden air — passive or 'secondhand' smoking — puts the breather at risk. Note the following:
- In 1986 the World Health Organization of the United Nations stated in a resolution that 'passive, enforced, or involuntary smoking violates the right to health of the nonsmoker, who must be protected against this noxious form of environmental pollution'.
- Tests on babies in homes where there is smoking show the presence of a nicotine derivative in their blood. Pneumonia and bronchitis are significantly increased in the infants of smoking parents.
- According to the 1984 report of the US surgeon general, the children of smokers are more prone to lung diseases than the children of nonsmokers.

Many communities have enacted laws that limit or forbid smoking in restaurants, lifts, and other public places.

Chewing tobacco, once the hallmark of the cinema cowboy, has recently become popular again, especially among children and young people. Dried tobacco, in shredded or brick form, is held in the mouth or chewed. The risk of cancer of the mouth is about four times as great in users of chewing tobacco as in nonusers.

S/N RATIO   *See* NOISE

SOARING   *See* GLIDER

SOFTWARE   *See* HARDWARE

SOLAR BATTERY   *See* BATTERY (CELL)

SOLAR ECLIPSE   *See* ECLIPSE

SOLAR ENERGY   *See* ENERGY RESOURCES

SOLAR FURNACE   *See* ENERGY RESOURCES

SOLAR HEATING

A means of converting sunlight into heat, usually for heating water.
See also ENERGY RESOURCES

SOLAR MASS   *See* STELLAR EVOLUTION

SOLAR PROMINENCE *See* SOLAR SYSTEM

SOLAR SYSTEM

A group of celestial objects moving in orbit around the sun (Latin *sol* sun). In addition to the sun and the earth, the group includes eight known planets and their more than 40 satellites, hundreds of known *comets*, thousands of *asteroids* (also called *minor planets* or *planetoids*), and countless trillions of *meteoroids* — pieces of matter from boulder size down to microscopic grains of dust.

Sun Our sun, at the centre of this assemblage, is a ball of hot gasses, about 1.4×10^6 kilometres (865,000 miles) in diameter. Its mass makes up more than 99.9% of the system. Thus its gravitational attraction (see GRAVITATION) is great enough to hold together the entire system.

The sun is a star of average temperature, about 15 million degrees Celsius in the interior, and about 6000°C at the surface — if a mass of seething gases can be said to have a surface. The outpouring of heat and light energy is caused by nuclear fusion, a process in which hydrogen, which makes up more than 90% of the sun's mass, is converted to helium. The energy, radiating through space, is the only significant source of heat and light for the solar system. At the present rate of conversion, the sun's supply of hydrogen fuel should suffice for about another 5000 million years of energy production.

In the intense nuclear-generated heat of the sun, all matter is gaseous. Nevertheless, the *photosphere* (Greek *phot*, light), a relatively dense layer of gases about 320 kilometres (200 miles) in thickness, is regarded as the bottom layer of the sun's atmosphere, where it joins the sun's 'surface'. The photosphere is the bright 'face of the sun'. Around it the faintly glowing pink *chromosphere* (Greek *chroma*, colour) extends to a height of about 9660 kilometres (6000 miles), and beyond that is the sun's *corona*, whose electrically charged particles, mainly hydrogen nuclei and electrons, hurtle hundreds of millions of kilometres into space. Some of the particles are caught and held by the earth's magnetic field (see SOLAR WIND; VAN ALLEN BELTS). The chromosphere and corona are too faint to see, except during a solar ECLIPSE, or at any time through a special type of telescope called a *coronagraph*. Here and there, similarly faint *solar prominences* extend many thousands of kilometres above the chromosphere.

Unlike the uniform rotation of the earth — every place on earth goes around, predictably, once daily — the rotation of the gaseous sun varies from about 25 days at the equator to 34 days at the poles. Some scientists have suggested that the uneven rotation of gases is

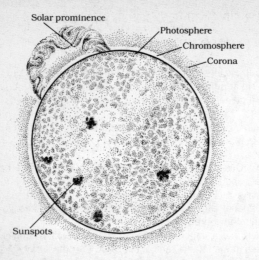

the cause of *sunspots*. They believe that uneven rotation distorts the sun's magnetic field, producing hot spots of intense magnetic strength in some places. The electrically charged gas particles are affected by magnetic forces; they encounter more resistance to movement over the hot spots than they do over the rest of the magnetic field, so fewer particles of gas move through these areas. There is less heat and less brightness there — in short, these are the cooler areas called sunspots. Actually they are bright, but they appear dark against the brilliance of the photosphere. A spot may be from 1000 to about 40,000 kilometres (600 to 25,000 miles) in diameter. Some spots appear, move across the photosphere as the sun rotates, and disappear. Others may pass around the edge of the sun and reappear later at the opposite edge.

The number of observed sunspots varies. There is a regular pattern, the *sunspot cycle*, with spots reaching a maximum every 11 years on the average. The pattern is duplicated in other forms of solar activity — the intensity of the streamers in the corona, for example, and the number of flamelike solar prominences seen at the sun's edges. These variations affect events on the earth. With an increase in solar activity, more charged particles reach the earth's atmosphere. They cause intensified AURORAS, they disrupt radio and TV reception, and there is evidence that they may influence our weather.

Planets The planets are large objects (the smallest to qualify in our solar system is Pluto, whose diameter is about 3220 kilometres/

2000 miles) revolving around the sun in elliptical orbits. Planets produce no light of their own and are seen only because they reflect sunlight. Eight of the nine known planets fit neatly into two groups:

- The four nearest the sun — Mercury, Venus, Earth, and Mars — consist mainly of rock and are called *terrestrial planets* (Latin *terra*, earth, land). They have atmospheres, with the exception of Mercury, whose extremely high temperature and low surface gravity preclude its holding on to the particles of gas that would make an atmosphere.
- The next four — Jupiter, Saturn, Uranus, and Neptune — are the *giant planets*, with diameters from four to ten times the diameter of the earth. All are composed largely of frozen gases, mainly hydrogen and helium.

The exception to the neat categories is Pluto, the outermost known planet and the most recently discovered (1930). It has been suggested that Pluto is not a planet but an escaped moon of Neptune; there is also speculation that it could be the brightest one of a group of small planets in the outer reaches of the solar system.

The numbers that describe sizes and distances in the solar system are more likely to astonish than to inform, as the chart on pages 276–77 confirms. But there are comparisions which help to make them manageable.

Size Let's reduce the sun's 1.4×10^6-kilometre (865,000-mile) diameter to 1 kilometre. On that scale, the largest planet, Jupiter (142,800 kilometres/88,700 miles in diameter), is reduced to less than 183 metres (600 feet) a little under the size of two football pitches end to end; the earth is less than the length of a tennis court; and Pluto is a bit bigger than a Ping-Pong table.

Distance The sun's light takes about 8 minutes to reach the earth. It takes more than 5 hours to reach Pluto. Astronomers and non-astronomers alike benefit from the system of *astronomical units (AU)*, in which small numbers are used to express the vast distances within the solar system. One AU is the average distance between the sun and the earth — 149,658,880 kilometres (92,955,832 miles). Two AUs are twice that distance, and so on. How far from the sun is Mars? 1.5 AU, Jupiter? 5.2 AU. Pluto? 39.5 AU. Clearly, this is much easier than juggling millions and billions when comparing distances of planets, comets, and other bodies from the sun. For the far greater distances outside our solar system, a far bigger unit is used (see LIGHT-YEAR).

How did the planets get their start? For centuries, two main lines of explanation have been offered, and there is still no definitive answer.

SOLAR SYSTEM

1. The *catastrophic theories* favour this scenario: the sun is struck by another star. Enormous masses of gas are thrown out into space, and some of them are pulled into orbit around the sun by its gravitation. The masses cool gradually, condensing eventually to form planets. In a variation on this theory, there is no collision, but the intruding star passes close enough to the sun to pull some of its gases out into space.
2. The *nebular theory* (Latin *nebula*, cloud) holds that the sun, like other stars, was formed by the congregation of vast clouds of gas and dust. Material that escaped the ingathering was drawn together by mutual gravitation into smaller clouds, which in time consolidated into planets.

Comets During most of its lifetime, a comet (Greek *kometes*, long-haired) is a chunk of frozen matter entombed in a cloud of dust and gases. That's a far cry from the long-tailed glowing ball the word *comet* brings to mind.

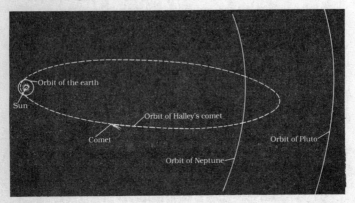

Planetary Data

	Diameter in kilometres/miles	Surface gravity (earth = 1)	Time for one rotation (in earth time)
Mercury	4878/3030	0.4	59 d
Venus	12,155/7550	0.89	244 d
Earth	12,756/7926	1.0	23 h 56 m
Mars	4217/6787	0.4	24 h 37 m
Jupiter	142,808/88,700	2.54	9 h 50 m
Saturn	120,430/74,880	1.07	10 h 40 m
Uranus	51,840/32,200	0.8	12–24 h
Neptune	49,500/30,750	1.2	15–20 h
Pluto	3220/2000	?	6 d 9 h

Note: m = minutes, h = hours, d = days, y = years

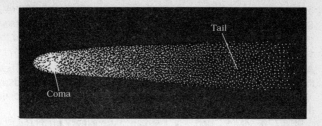

Comets lead a kind of double life. Like planets, they are a part of the solar system, moving in an orbit around the sun. But the orbits of most comets are tremendously elongated, with the sun near one end and the far reaches of space at the other. For example, Halley's comet, the best-known member of the family, comes within 87 million kilometres/54 million miles (0.587 AU) of the sun, then races away to a distance of about 5300 million kilometres/3300 million miles (35.33 AU), to return approximately 76 years later, again and again.

On its inward bound trip, a comet begins to warm up. Its solid centre, called the *nucleus*, at most a few kilometres wide, begins to melt. Electrically charged gases are released, forming a head, or *coma*, 16,000 kilometres (about 10,000 miles) or more in diameter. Particles from the sun cause the atoms and molecules of the gases to glow; they also drive the gases away from the coma, forming a glowing tail that may be many millions of kilometres long, pointing away from the sun. After the comet has rounded the sun and begun its outbound trip, it cools and returns to the frozen condition.

A comet loses part of its substance on each return. Why then don't we run out of comets? In 1950 Jan H. Oort, a Dutch astronomer, proposed the theory that the material for billions of potential comets lies in a sphere (now called the *Oort comet cloud* or *Oort cloud*)

Time for one revolution (in earth time)	Number of satellites	Average distance from sun	
		In millions of kilometres/miles	In AU (earth = 1)
88 d	0	57.91/35.97	0.387
225 d	0	108.25/67.24	0.723
365.26 d	1	149.66/92.96	1.0
687 d	2	228.0/141.6	1.5237
11.8 y	18	778.6/483.6	5.2028
29.5 y	13	1427.6/886.7	9.5388
84 y	5	2872.4/1784.1	19.1914
165 y	2	4499.1/2794.5	30.0611
248 y	1	5916.1/3674.6	39.5294

around the solar system at a distance of at least 50,000 AU. Sometimes a change in gravitation, perhaps caused by the passage of a star, sets a comet on a course that eventually brings it into the sun's gravitational field, and it is pulled into orbit around the sun.

Asteroids Asteroids (Greek *asteroeides*, starlike, which is how they appear in a telescope) are more properly called planetoids or minor planets, because, like tiny planets, they move in orbits around the sun.

In 1772 Johann Bode, a German astronomer, predicted the existence of an undiscovered planet in the large gap between the orbits of Mars and Jupiter. His announcement was partly responsible for the organization of the 'celestial police', a group of astronomers dedicated to finding the planet with their telescopes. No full-sized planet was found, but in 1801 Giuseppe Piazzi, an Italian astronomer (who was not one of the policemen), spotted a new object in the area. It was named Ceres, after the Roman goddess of agriculture (Demeter in the Greek pantheon).

Ceres, with a diameter of approximately 1000 kilometres (over 600 miles), is the largest of the thousands of known planetoids, which range downwards to boulder size. Their combined total mass is estimated at less than one-thousandth of the earth's mass. Nearly all the minor planets travel in the *asteroid belt*, between the orbits of Mars and Jupiter. Occasionally minor planets collide, and pieces of debris, drawn by the earth's gravitation, flash through our atmosphere as *meteors*.

Meteors Also called (picturesquely but wrongly) a 'shooting star' or a 'falling star', a meteor is seen as a momentary streak of light through the sky. Especially bright meteors are called *fireballs*, and a fireball that splits is a *bolide* (Greek *bolis*, missile).

A meteor begins as a *meteoroid* in interplanetary space. It is a piece of stony or metallic material, most likely from an asteroid or a comet. Meteoroids may range in size from smaller than a pinhead to bigger than a house. As the earth sweeps along in its orbit around the sun, its gravitational attraction captures nearby meteoroids, pulling them down at speeds of up to 56 kilometres (35 miles) per second. Friction with the atmosphere about 95 kilometres (60 miles) above the earth heats the meteoroid and the air around it to a temperature of over 2000°C (3600°F). In the intense heat, atoms and molecules of air split and recombine, producing the glow of light. Small meteoroids are completely vaporized. Large chunks of material may burn away only partially; the chunk that remains, falling on the earth, is called a *meteorite*.

See also BIG BANG; COSMOLOGY; STELLAR EVOLUTION

SOLAR WIND

A stream of electrically charged particles issuing from the sun's outer atmosphere, called the *corona*. The particles, mainly hydrogen nuclei and electrons, are driven out in all directions by the sun's intense heat; they race past the earth at speeds of over 320 kilometres (200 miles) per second and probably continue on far past Pluto, the outermost known planet.

Some of the particles, trapped by the earth's magnetic field, are held high above the earth in rings known as the VAN ALLEN BELTS. The particles are the cause of AURORAS. They also disturb radio and TV reception because they carry electric charges that emit radio waves. You may have experienced a similar effect on a car radio when another car with defective spark plugs drew up alongside, causing crackles and sputters on your programme.

SOLID   *See* MATTER

SOLID-STATE PHYSICS   *See* PHYSICS

SOMATREM   *See* GENETIC ENGINEERING

SONAR

Acronym for *s*ound *n*avigation *a*nd *r*anging, a technique and apparatus for determining the location of an object by reflected sound waves. Also called *echolocation*, it was first invented by bats, who neglected to have it patented. Bats flying at night emit ultrasonic squeaks, inaudible to humans but not to bats. The squeaks are reflected from solid objects such as cave walls or flying insects and return as echoes to the bat's ears. By timing the interval between the emission and return of squeaks, a bat computes its distance from the object. And by comparing the loudness detected by the left ear and the right, it computes the angular direction of that object.

The sonar principle is used to determine the depth of shallow bodies of water and to locate schools of fish, underwater submarines, mines, wrecks, and other obstacles. In *active sonar*, pulses of high-frequency (high-pitched) sound are beamed downwards and at angles from the bottom of a ship. The echoes are received by an apparatus that measures the time interval, then computes the distance and direction of the reflecting object. This information is shown on a dial or plotted automatically on a chart. *Passive sonar*

SONOGRAM

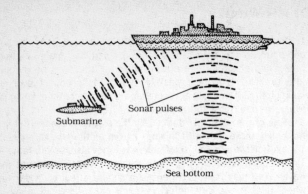

does not send out sounds; it detects sounds made by submarine engines or other sound-producing objects.

See also RADAR.

SONOGRAM See ULTRASONICS

SORBITOL See SWEETENING AGENTS

SOUND ENERGY See ENERGY

SOUND RECORDING

A method of converting sound waves into permanent form, to be stored and later reproduced. Before we examine sound recording, let's have a graphical look at sound itself.

Sound results from the audible vibration of some object. On the left of the diagram is a magnified view of a groove cut into a gramophone record, made by the vibration of a tuning fork sounding middle C. The shape is smooth, simple, and even. Mathematically, it is a *sine wave*. On the right is the same sine wave (A) drawn as a graph. Graphs B, C, D, and E show the shape of grooves cut by the sound of a flute, a violin, a clarinet, and a certain soprano singing 'eeeeeee'.

The basic (fundamental) repeating sine wave can still be picked out, embellished by variations called *overtones*. It is the overtones that enable us to distinguish one voice or instrument from another.

How were the grooves in the record made? The earliest commercial sound recorder was invented by Thomas A. Edison (1847–1931). One played or spoke, as loudly as possible, into the wide end of a horn. At the small end, a disc was made to vibrate by the sounds.

SOUND RECORDING

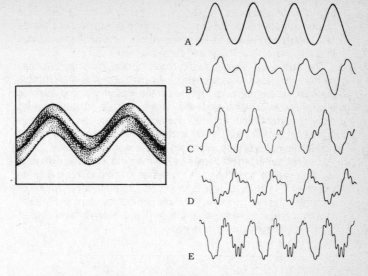

Attached to the disc was a short needle whose point pressed against a cylinder of soft wax. The vibrating needle dug a groove into the wax as the cylinder rotated.

If you were given the choice of carving one of the grooves shown earlier, you would no doubt choose A because its smooth, regular curves are comparatively easy to cut. It's more difficult to cut the flute's wave-form, with the overtones that make it sound like a flute. As for the violin and the human voice, with their subtleties of

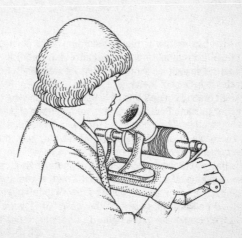

overtones on overtones on overtones, no early recording technique could faithfully reproduce them. The needle point could not vibrate fast enough in the wax to incise the minute high-pitched overtones that give character to music and the human voice.

The next great step in sound recording was the magnetic-tape method. A plastic tape is coated with particles of a magnetizable material such as iron oxide. Millions of the randomly placed particles occupy each centimetre of the tape's surface. Electromagnets carrying currents from a microphone jockey the particles into a series of magnetized stripes, one for each vibration. Taping is an almost perfect recording system, especially when enhanced by Dolby.

It is not quite perfect because each magnetic particle, light as it is, does weigh something. It possesses a certain resistance to movement, *inertia*, which prevents it from obeying fully the extremely feeble commands of the tiniest overtones. Is there anything with less inertia than the particles? Yes — beams of laser light, which have virtually no inertia.

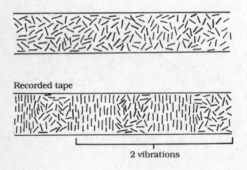

Recorded tape

2 vibrations

Laser light has another virtue for recording. It can be focused to a point 0.000025 millimetre (one-millionth of an inch) in diameter. At that focus, the light is so intensely concentrated that it can punch a tiny crater in a sheet of metal.

Now what exactly does all this have to do with recording the golden overtones of Luciano Pavarotti? One more sidetrack and we can put everything together. The sidetrack is the concept of *digital scanning*. This is the basis of digital recording, which may someday entirely displace present-day recording methods.

Here's the flute wave again, drawn on a sheet of graph paper with numbers along the edge. You could direct a partner over the phone in making a copy of the wave on another sheet of graph paper. Start at the left and call out the height of the wave (0). Then call out its height at the next square to the right (3). Continue to find the height of the wave (scanning it) and calling out, while your partner makes dots at the corresponding places on his or her paper. After joining the dots together with a line, your partner will have a fair copy of the original flute wave graph. To make a more exact copy, use graph paper with smaller squares. The smaller the squares, the greater the amount of information (number of scans) and the more accurate the copy.

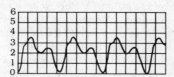

At last we get to digital recording and to Pavarotti, who has been holding his breath all this time. As he sings into the microphone, an electronic scanner measures the height of each wave several thousand times, one wave after another. The scanned information is passed along in the form of electrical pulses — binary digits, ones and zeros — to a recording head. The head is a laser lamp focused to a pinpoint on a revolving metal disc. The lamp is controlled by the electrical pulses from the scanner. A one bit causes the lamp to emit a burst of light, which digs a tiny crater in the surface of the disc. A zero bit causes no response from the lamp, so a blank space is left on the disc. Wave after wave, the disc becomes studded with millions of craters and spaces in a spiral track. This is the master disc, from which several thousand copies — compact discs are made. These copies can be played back only on special digital playback equipment.

The digital players scans the surface of the record by means of a low-power laser beam. It transmits the information (e.g., crater, space, space, crater) to a computer that converts the bits into electric currents of varying strengths. These currents are fed into a loudspeaker, producing sound waves almost identical with Pavarotti's original waves.

In making and playing ordinary records, there is always contact between some part of the equipment and the record. In digital recording, because the only link is the laser beam, the scratches, pops, and hisses of ordinary recording are eliminated.

Digital systems can be used to store and reproduce many kinds of information. Videotapes, for example, reproduce the sounds and pictures of films, because pictures can be scanned and recorded as readily as sound waves.

See also DOLBY NOISE REDUCTION SYSTEMS.

SPACE, RELATIVE See RELATIVITY

SPACE TRAVEL

This piece is addressed to the mature reader — the person handicapped by the possession of common sense. Such sense evokes a wince at the sight (news photo, TV) of an astronaut jockeying a strayed satellite into tether, with nothing but vertiginous space between himself or herself and the blue and brown cloud-flecked earth far, far below. Why isn't the astronaut falling? Common sense, arising from common experience, points to such an outcome. Or are there special laws of physics that operate out in space but not close to the earth? The less mature reader, nurtured on space comics, science fiction, and the like, may readily inform you that there is no gravity in space (wrong) or that centrifugal force is the saviour (not so). Therefore, to you, mature reader, we offer a brief compendium on space travel, pared to the bone. In fact, three bones:

1. What gets the spaceship off the ground and out into space?
2. What keeps it up there after the engines stop (as they do after about 10 minutes)?
3. What brings it back to earth safely?

We would recommend first a reading of the briefly and lucidly written entry on NEWTON'S LAWS OF MOTION.

Now, what gets the spaceship off the ground and out into space? This goal was visualized by Jules Verne in 1865 in his novel *From the Earth to the Moon*. Verne proposed a huge cannon whose projectile was a bullet-shaped cabin with all the comforts of home for its occupants — three humans, a dog, and a rooster. Consider a few pros and cons:

With gravity clutching at it inexorably, the projectile begins slowing down from the moment it leaves the cannon. That deceleration would bring the projectile, sooner or later, to a slowdown, halt, and return to earth.

However, the pull of the earth's gravity weakens with distance from the earth (see INVERSE SQUARE LAW). Suppose we start off at such a great speed that even though we're slowing down, we still have enough reserve speed to take us out to a distance where the earth's gravitational pull is much weaker. Then another force (hold on for a moment) could take over.

How to achieve that great initial speed? Verne proposed an enormous amount of propellant for the cannon: 400,000 pounds

(181,440 kilograms) of gunpowder, blasting the projectile through a 900-foot (274.5-metre) cannon.

However, the initial effect of such a supercannon would be smashing indeed. Living matter could not withstand the crush of such powerful acceleration from dead still to full speed within a few seconds.

However, why must we deliver all that force in a single bang? It's necessary only if the bang is to be delivered by gunpowder inside the cannon. But suppose we attach the gunpowder to the rear of the projectile, arranged so it fires backwards in a steady stream. Thus we can build up the speed gradually and safely.

That idea has probably aroused a commonsense hesitation in you, to wit: what happened to the cannon? To which a scientist would reply, 'Who needs it?' The energy to drive the projectile is in the propellant — the gunpowder or one of the various rocket fuels. The cannon is merely a convenient container. Consider this humble prototype of a spaceship.

The air pressure in a balloon is pressing outwards in all directions. Yet, although the balloon is free to go, it does not, because all the interior pressures are equal: forward, backward, and sideward, the pressures are balanced. (Newton's first law, first half: a body at rest will remain at rest unless acted upon by a force.) Opposite-acting equal forces don't count: $1 - 1 = 0$.

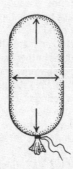

But suppose we loosen the string at the neck of the balloon. Now we have an unbalanced situation: the forward pressure pushes forwards against the balloon; the rearward pressure simply escapes. As long as there is inside pressure, the balloon will be pushed forwards, faster and faster, because there's no rearward pressure to balance the forward pressure.

In the same way, the rocket engines in a spaceship produce immense quantities of high-pressure gas, thrusting forwards, back-

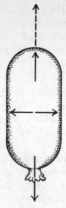

wards, and sidewards. The forward pressure (upward, at launch position) does the basic moving job. The spectacular downwards-plunging plumes of flame-lit gas make a lovely contribution to TV, but the really hot news is being made inside, up against the forward ends of the engines.

Bravo! The upward force exerted by the burning fuel increases until it equals and exceeds the downward pull of gravity. We achieve *lift-off*. We accelerate to the charmed number of 11.2 kilometres per second (7 miles per second), *escape velocity*. At this speed, even though gravity still pulls us towards the earth, it can no longer slow us to a stop and turn us around.

What keeps the spaceship up there after the engines stop? Headed in the right direction, with no further power needed, we can just keep going, to the end of time. (Newton's first law, second half: A body in motion will remain in motion, in a straight line, *unless a force acts on it*.) Space is full of forces acting on our spaceship and on everything else. Every object in the universe is exerting its gravitation on us and on all other objects. However, their pulls just aren't strong enough to bother us significantly during lift-off and escape. But we'd better watch where we're going.

Suppose we're not quite so adventuresome, and a few orbits of the earth will suffice. No problem: just stay below escape velocity. Depending on how much below, we can stay close to home and call it a day after a few hours or go farther and thereby stay longer, or even station ourselves in a *geostationary (geosynchronous)* orbit, 35,900 kilometres (22,300 miles) high, where our velocity of nearly 11,270 kilometres (7000 miles) per hour keeps us continually over the same spot above the equator on the rotating earth.

This is the orbit where *communications satellites (comsats)* such as the *Westar* (US), *Anik* (Canada), and *Palapa* (Indonesia) are

SPACE TRAVEL

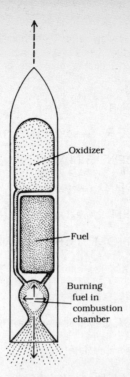

parked. These serve as amplifiers and reflectors of radio, television, and other electronic-wave man-made information.

What brings the spaceship back to earth safely? Any object in earth orbit is slowed down by frictional forces, such as contact with the atmosphere, thin as it is. It may take days, years, or centuries, but eventually it happens — the orbiting object is slowed down; it spirals lower and lower, reaching denser atmosphere, where increased friction raises its temperature higher and higher, usually to the point of destruction. If astronauts are aboard, obviously provision must be made for a gentler homecoming. This is done in two steps:

1. The spaceship's outer skin has a heat shield, a covering of special ceramic tiles, to be sacrificed in the process of *ablation* (Latin *ablatus*, carried away). These tiles become red-hot, melt, and vaporize into the atmosphere, thus dissipating the heat that would otherwise scorch the spaceship and its contents.
2. During the final part of its descent, the spaceship changes character. It was, all during its flight, a ballistic object (Greek *ballein*, to throw). It was thrown into space by its engines, then

left to make its way by inertia (Latin *iners*, idle). Now it turns into a glider, by virtue of possessing wings, control surfaces, and, most desirably, a human pilot who brings it to a welcome touchdown.

SPECIES See EVOLUTION

SPECTROMETER See DOPPLER EFFECT

SPECTROSCOPIC BINARIES See BINARY STAR

SPECTROSCOPY See ANALYSIS

SPEED OF LIGHT See RELATIVITY

SPHYGMOMANOMETER See BLOOD PRESSURE

SPIRAL GALAXY See GALAXY

SSC See PARTICLE ACCELERATOR

STAPHYLOCOCCUS See ANTIBIOTICS

STAR See STELLAR EVOLUTION

STAR MAGNITUDE See MAGNITUDE

STAR WARS See STRATEGIC DEFENSE INITIATIVE

STATIC ELECTRICITY See FIELD THEORY

STATIONARY FRONT See AIR MASS ANALYSIS

STATOR See GENERATORS AND MOTORS, ELECTRIC

STEEL See ALLOY

STEEL ROLLING See MANUFACTURING PROCESSES

STELLARATOR See NUCLEAR ENERGY

STELLAR EVOLUTION

Stars are not eternal, but they do shine for a very long time — thousands of millions of years, usually. Yet scientists can trace the evolution of stars — the changes they undergo — from birth to death. The trick lies in studying stars in a range of stages, as one might study aging in humans by observing many people in all stages of life, from infants to the very old. We'll use these stages for our quick look at stellar evolution: birth, midlife, decline, and death.

Birth Stars begin their lives in 'empty' space, which isn't really empty. It contains the *interstellar medium*, widely dispersed particles that are about 99% gas (mainly hydrogen) and 1% solid ('dust'). In some areas, where the density of the medium is a bit greater, clouds of particles may be drawn together by mutual gravitation. Other clouds may form from materials ejected by existing stars.

A cloud grows by attracting more molecules of gas and dust; the consequent increase in its gravity pulls the molecules together more and more tightly. The cloud continues to attract material, and continues to collapse, denser and denser. In the crushingly packed mass, the normal heat-producing vibratory motion of the molecules — their collisions and rebounds — is vastly increased.

The heat raises the temperature within the mass (now elevated to the rank of *protostar*, from Greek *proto*, before) to millions of degrees. It also starts up the process of *nuclear fusion*, and a glowing star is born. The fuel that feeds the nuclear fire is hydrogen. It is converted to helium, and enormous amounts of energy are released in the form of heat, light, and other kinds of radiation.

EVOLUTION OF STARS

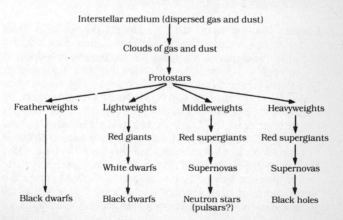

Midlife The intense heat in the star's core builds up a pressure that forces some of the gas outwards, closer to the surface. The remaining gas cools somewhat, and a balance of forces is reached: the outward pressure, caused by the heat, versus the inward tug, caused by gravitation. The balance prevails while enough hydrogen fuel remains in the core. Within the star nearest to us, the sun, this balanced state of affairs is only halfway through its expected 10,000-million year stretch. (Pessimists may prefer '*already* halfway through'.)

Decline As the hydrogen supply dwindles, the core begins to cool and the outward pressure lessens. The result is that the balance of forces tilts in favour of gravitation, there is a new series of heat-generating gravitational collapses in the core, and the outward pressure again builds up. At the same time, hydrogen in the outer layers of the star continues burning, advancing the star's surface outwards again and enlarging it enormously. When our sun reaches this stage, it will have become a *red giant* star, big enough to engulf Mercury and Venus in their orbits and reaching almost to the earth. Giant indeed, but why red?

A star's colour is the key to its surface temperature. Like a bar of iron removed from a blacksmith's forge and cooling, the colours range from blue-white, for the hottest, through white, yellow, and orange, to red for the coolest. All stars are coloured, although the eye can detect this only in the brightest ones, such as Sirius, a white star whose surface is about 10,000°C; Capella, a yellow star like the sun, which is about 5000°C; and Antares, which is red and under 3500°C. When a star's size increases, its heat and light radiate from a surface that is very much enlarged and in a relatively cool stage. Hence the term *red* giant.

Death The length of a star's life and the manner of its death depend on how massive it was to begin with — how much matter it contained. Like boxers, stars can be conveniently classified as featherweights, lightweights, middleweights, and heavyweights. Let's take them slightly out of order, beginning with the lightweights.

Most stars, including the sun, belong among the lightweights. On a scale where the sun's mass (the *solar mass*) is 1, lightweights range from about 0.1 to 4.0. Such stars, toward the end of their red giant stage, may go through a third series of gravitational collapses that restart nuclear fusion in the core; this time the helium that resulted from hydrogen burning is burned in its turn and converted to carbon. Pressure outward builds up again, sufficient to produce a second red giant stage and even to blow away a large part of the outermost gases. A star may lose more than half its mass in this way.

What remains of the one-time red giant is a *white dwarf* — a ball of gas no larger than the earth and about 300,000 times as dense. A teaspoon of a white dwarf, brought to earth, would weigh several tons. The dwarf, incapable of further nuclear reactions, cools slowly over hundreds of millions of years. The star that was once the centre of unimaginable heat and violence comes to its end as a *black dwarf* — an invisible ball of extremely dense gases in space.

Featherweights never reach true stardom. Their solar mass is too small — much less than 0.1 — to generate the gravitational pressure and consequent high temperature needed to trigger the nuclear fusion process. The failed star cools, ending as a black dwarf.

Middleweights are stars with a solar mass between 4 and 8.

Heavyweights have a solar mass above 8.

> My candle burns at both ends;
> It will not last the night;
> But, ah, my foes, and, oh, my friends—
> It gives a lovely light!

The words of the American poet Edna St Vincent Millay could have been meant for the middleweight and heavyweight stars. Their hydrogen is burned at a much faster rate than in lightweights. They produce enormous heat and spectacular brilliance, but at a price: a star ten times as massive as the sun lives only about a thousandth as long. (Still, 10 million years ...) All stars have much the same history in their early stages, up to the red giant stage, but after that a massive star, instead of collapsing like a lightweight, continues to expand. It reaches a diameter hundreds of times the diameter of the sun, justifying its name of *red supergiant*. Betelgeuse, an easily seen star in the constellation of Orion, is a red supergiant.

Supernovas Within the core of a red supergiant, each burnout of nuclear fuel is followed by another heat-generating gravitational collapse that restarts the fusion process. The repeated burnings convert lighter elements to heavier ones: hydrogen to helium to carbon to magnesium to iron. At that point, for reasons not fully understood, there is an especially powerful collapse and rebound of the core at the same time that the outer layers of gas collapse onto it. The resulting explosion may destroy the star completely or blow most of it away, leaving only the crushed mass of the core.

The burst of light accompanying the explosion may last for weeks and be 100 million times as bright as the star was. Another product of the explosion may be *cosmic rays*. This catastrophic period is the star's *supernova* stage (Latin *novus*, new); although the star is very old, its extraordinary brilliance is new). A NOVA is a much less dramatic flare-up by a star, arising from another cause.

Supernovas are extremely rare. In AD 1054, Chinese astronomers observed a 'guest star' bright enough to be seen in the daytime. Today the remnants of this supernova's blown-away outer layers are barely visible through binoculars as a faint, cloudlike smudge, the *Crab Nebula*. A dim star near the centre of the nebula, visible only in a powerful telescope, is probably the core of the original star.

In February 1987 there appeared an unusual supernova that promised much new information, for two reasons. It was in the Large Magellanic Cloud, the galaxy closest to the Milky Way. Also, it was the first supernova whose progenitor star had been observed and photographed (many times, in fact) *before* the explosion. Scientists believe its mass was 15 to 20 times that of the sun.

End of a Massive Star Evidence collected in the last half-century points to two possibilities for the last stage of a supernova core:

Neutron Stars: In the atoms of 'ordinary' matter, the protons, which are surrounded by clouds of electrons, repel one another; 'ordinary' matter, then, is mainly empty space. But the gravitational collapse of a middleweight star, having a solar mass of 4 to 8 creates a pressure great enough to crush the electrons right into the protons, creating *neutrons*, which do *not* repel one another. The matter in such a *neutron star* would be packed so densely that a teaspoonful of it would weigh several million tons if brought to earth! Put another way, our sun, crushed down to this density, would be less than 16 kilometres (10 miles) in diameter. It now seems almost certain that neutron stars are PULSARS.

Black Holes: According to theory, the collapse of a heavyweight star, one whose solar mass is greater than 8, would produce a core so dense, with a gravity so powerful, that it would suck in all nearby matter. It would also seize all radiation, including its own light, in a fierce grip and thus be invisible. Nevertheless, several black holes have been observed indirectly, by satellite-mounted instruments designed to detect X-rays. Binary stars consist of two or more stars held in orbit by their mutual gravitation. A visible star may be the partner of a black hole and may be close enough for some of its gases to be pulled towards the black hole. In such a case the heated matter radiates X-rays in all directions as it falls towards the black hole, providing a means of indirect observation.

See also NUCLEAR ENERGY.

STEP-DOWN TRANSFORMER   *See* TRANSFORMER

STEP-UP TRANSFORMER   *See* TRANSFORMER

STEREOPHONIC SOUND   *See* QUADRAPHONICS

STEROIDS

A large class of chemical compounds occurring naturally in plants and animals and also made synthetically. The natural steroids include the *steroid hormones*, which can be further divided into the *sex hormones* and *adrenocortical hormones* (see below); *sterols* (CHOLESTEROL and related compounds); and the *bile acids*, chemicals important in the digestion of fats.

Sex hormones These fall into three categories:

Androgens Male sex hormones, such as *testosterone*, produced mainly by the testes. They are responsible for the development and maintenance of the male sex organs and sexual characteristics.

Oestrogens Female sex hormones produced by the ovaries. They have a similar role in females as the androgens in the male.

Progesterone A hormone released to prepare for and maintain pregnancy.

Adrenocortical hormones These are hormones produced by the cortex (outer layer) of the *adrenal glands* at the top of each kidney. Like the sex hormones they are all derived from cholesterol:

Glucocorticoids Adrenocortical steroids, such as *cortisone* and *hydrocortisone*, concerned principally with the breakdown of protein and formation of glucose, and with the use of carbohydrates in the body. Medically, they are used in treating rheumatoid ARTHRITIS, allergies, and skin disorders.

Mineralocorticoids *Aldosterone* and related hormones that control water and salt balance in the body.

Synthetic steroids These are man-made drugs which have actions like the steroid hormones. The best known are:

- *Anti-inflammatory agents* derivatives of the glucocorticoids such as prednisolone and dexamethasone which are used to treat rheumatoid ARTHRITIS.
- *Oral contraceptives*, an oestrogen, a progesterone derivative, or, most commonly, a combination of both.
- *Anabolic steroids*.

Anabolic steroids The most familiar of these is testosterone which has both androgenic and *anabolic* (from Greek *anabole*, building up) properties. Some synthetic anabolic agents are nandrolone, stanozolol, and methandienone. In 'legitimate' use, anabolic steroids have been used to treat cases of wasting, *osteoporosis* (porous brittle bones), and some types of anaemia.

Their most publicized use, however, is by body-builders and athletes. Anabolic steroids, taken during training, have the effect of increasing muscle bulk and strength and body weight. Although banned by all sports authorities it is believed that they are widely used to enhance athletic performance, a situation which has caused increasing concern. Highly sophisticated drug-testing techniques have been developed that can detect the minutest concentrations of anabolic agents in urine. *Masking agents*, which are sometimes taken to try and disguise use of anabolic steroids, can also be detected by these techniques. Masking agents are substances that can alter the way in which steroids are excreted in the urine; a commonly used one is *probenecid*, a drug used in treating gout.

Apart from the ethical considerations, the side-effects of anabolic steroids make their use of questionable value. They are known to increase aggression, produce abnormalities of the genitals, and increase the risk of both liver and heart disease. Some masculinization may occur if taken by women.

STEROLS   See STEROIDS

STOMATA   See TRANSPIRATION

STORAGE BATTERY   See BATTERY (CELL)

STORAGE RINGS   See PARTICLE ACCELERATOR

STRATEGIC DEFENSE INITIATIVE (SDI)

A high-technology military programme of the United States, begun in 1983. Commonly called Star Wars, SDI differs from conventional warfare (including 'conventional' nuclear weapons) in its objective: to recognize, intercept, and destroy enemy missiles while they are *in flight*, preferably right after they have been launched. Among various technologies being researched are LASER weapons. Here the purpose is to concentrate powerful beams of laser-light energy into a very small space, thus creating an enormously high temperature, sufficient to shatter or disable a missile.

The principle has already achieved numerous successes. For example, in factories, thick sheets of steel are cut by laser beams at a distance of several centimetres. Not quite the same, you might say, as pinpointing a laser beam onto a bomb about the size of a rowing boat, several thousand kilometres away, and flying at supersonic

STRATEGIC DEFENSE INITIATIVE (SDI)

speed, and then keeping the beam fixed on a coin-sized area of the bomb long enough to achieve the required lethal temperature. Here are some approaches to this problem.

Mount a laser apparatus on a satellite in permanent orbit, aimed and controlled from a ground station. Or mount a mirror on a satellite, to reflect a laser beam from a ground station.

There are a few stumbling blocks. To achieve its temperature-raising effect, the laser beam must retain its narrowness; the least bit of spreading reduces the effect considerably (see INVERSE SQUARE LAW). Yet the atmosphere does scatter laser light a slight amount. Also, powerful as it is, a laser beam is light and can be reflected by any shiny surface, such as a mirror-coated missile.

Since the problem is loss of power (because the laser beam is weakened along the way by spreading, scattering, and reflection), why not start off with a superpowerful beam so that there's still enough punch left, after losses, to do its work? This is the idea behind *nuclear-powered lasers*. The force of a nuclear explosion would be converted instantaneously into a laser beam of unbelievable intensity.

Another Star Wars proposal is something called *smart rocks*. These projectiles, each about the size of a loaf of bread, would be launched from satellites in permanent orbit. Filled with electronic apparatus to latch on to the emanations (infrared waves, radio reflections, etc.) of an approaching nuclear bomb from several thousand kilometres away, this self-guided bullet would direct itself into collision course with the bomb.

Another part of the strategy is *particle beams*. These are composed of electrons, protons, and/or neutrons, whirled around and around in an apparatus that accelerates them almost to the speed of light and then hurls them. The apparatus is, in effect, a modified PARTICLE ACCELERATOR, mounted on a satellite. The particles are almost weightless, so their destructive power would come not from simple collision but from electrical and electronic damage. Since modern warfare is highly computerized, such beams, if successful, could disable an enemy's guidance systems.

Obviously, neither side could allow its expensive hardware to become helpless targets exposed in space, so countermeasures would be set up, and counter-countermeasures, ad nauseam. For example, a satellite-mounted particle accelerator would be vulnerable to antisatellite missiles and would need to be protected by anti-antisatellite missile satellites. And antimissile satellites could be fooled by decoy (blank) missiles into releasing their destructive force uselessly, thus allowing real missiles to get through to their unprotected targets.

In 1987 a committee of the American Physical Society, the

nation's largest association of physicists, concluded that it was 'highly questionable' that a space-based Star Wars system could survive an attack. One of the committee members put it this way: 'I am 99.9% sure it won't work'.

STREAMLINING

Shaping an object, such as an aeroplane or a ship's hull, to reduce its resistance to motion through a fluid such as air or water.
See also AERODYNAMICS.

STRIKE-SLIP MOTION *See* PLATE TECTONICS

STROKE

Brain damage resulting from an inadequate supply of blood to the brain. About 125,000 cases of stroke including 70,000 deaths, are recorded annually in the United Kingdom. Also called *cerebral apoplexy* and *cerebrovascular accident*, stroke may be caused by any of these events:

- Rupture of an artery in or around the brain, with loss of blood (haemorrhage).
- Blocking of an artery by an embolus – an object normally not in an artery, such as a blood clot or bubbles of air.
- Narrowing (stenosis) of an artery by deposits of fatty material on its inner wall; this is especially frequent in people with high blood pressure (*see* ARTERIOSCLEROSIS).

The effects of stroke vary considerably, from mild impairment of some body function to death, depending on the part of the brain involved and the extent of the damage. Speech and vision are often affected; muscular movement may be weakened to the point of paralysis. The work of a damaged area of the brain is sometimes partially taken over by another area. Speech therapy and physiotherapy may help in partial restoration of impaired functions.

A type of stroke with relatively mild, brief effects is the transient ischaemic attack (TIA), in which there is a temporary shortage of blood supply in a small area of the brain. The symptoms of such a ministroke may include vision problems, loss of feeling in a part of the body, and inability to speak or understand speech. The symptoms disappear in minutes, or in hours at most; however, TIAs may be a sign of increased likelihood of future strokes.

Recently developed procedures may improve the outlook for people who have had a stroke. One of these methods resembles coronary bypass surgery, in which a healthy blood vessel from another part of the body is used to shunt blood around a faulty vessel. In another method, blocked vessels are cleared by drugs or by laser beams. In an experimental method resembling KIDNEY DIALYSIS, the fluid that surrounds the brain is used as a vehicle to bring oxygen to the brain and to remove its waste products.

STRONG NUCLEAR FORCE

One of the four known fundamental forces of the universe; also known as the *strong interaction*. Its influence is limited to the atomic nucleus.

See also GRAND UNIFIED THEORY.

STRUCTURAL CHEMISTRY *See* CHEMISTRY

STRUCTURAL FORMULA *See* PROTEINS AND LIFE

STRUGGLE FOR EXISTENCE *See* EVOLUTION

SUBDUCTION *See* PLATE TECTONICS

SUBSTRATE (OF ELECTRONIC CHIP) *See* CHIP

SUN *See* SOLAR SYSTEM

SUNSPOT *See* SOLAR SYSTEM

SUNSPOT CYCLE *See* SOLAR SYSTEM

SUPERCONDUCTING SUPER COLLIDER
See PARTICLE ACCELERATOR

SUPERCONDUCTIVITY

An unusual condition in which certain substances lose all resistance to the flow of an electric current. If such a substance is connected to an electric source and then disconnected, the current within the substance continues to flow, on and on. This sounds like something for nothing, and the power companies had better watch out — that is, until the 'unusual condition' is examined. The substances (lead, tin, mercury, some other elements, and many compounds) must be cooled down in liquid helium to within a few degrees of ABSOLUTE ZERO (−273.15 °C or −459.7°F). This rules out superconductivity for ordinary use, where the savings on your electric bill would be so far more than offset by the expense of reaching and maintaining such a low temperature.

However, scientists have recently developed compounds that become superconductive at temperatures high enough to allow the use of liquid nitrogen, which costs about $1/50$ as much as helium. Using compounds based on thallium, superconductivity has been reached at temperatures of −150°C (−238°F), and temperatures as high as −95°C (−139°F) are predicted. Physicists are interested in superconductivity for many reasons, these among them:

1. A computer circuit consumes a tiny amount of electricity and generates a tiny amount of heat as it operates. In a large computer, millions of circuits generate heat that must be dissipated, or the circuits will break down. Superconductive circuits using virtually no current would generate virtually no heat. This would permit more circuits to be packed into a given space, in turn allowing faster operation of the computer.
2. PARTICLE ACCELERATORS use tremendous jolts of electric current, enough to light a small city. The current is used mainly to power hundreds of huge electromagnets. Under conditions of superconductivity, a given amount of power would greatly increase the amount of magnetism produced. Superconducting magnets would also allow the designing of smaller, more efficient GENERATORS AND MOTORS and of high-speed trains that float above their tracks, levitated by the magnets.
3. Electricity is transmitted from generators to users through metal wires. Some of the energy is wasted as the current flows, mainly as heat that warms the wires and the surroundings. Now scientists have produced flexible ceramic filaments, less than 0.25 millimetre ($1/100$ inch) thick, that become superconductive at a temperature of −196°C (−321°F), allowing the use of the relatively inexpensive liquid nitrogen. If thicker flexible filaments could be made, they might transmit electricity without heat loss.

SUPERCOOLING See PARTICLE ACCELERATOR

SUPERNOVA See STELLAR EVOLUTION

SUPERSYMMETRY (SUSY)

A number of theories leading toward the GRAND UNIFIED THEORY, which is expected to encompass all four known fundamental forces of the universe in one description.

SUPRESSOR CELLS See IMMUNE SYSTEM

SURVEYING, LASERS IN See LASER

SURVIVAL OF THE FITTEST See EVOLUTION

SUSY See SUPERSYMMETRY

SWEETENING AGENTS

Substances used to sweeten foods and beverages. The most common is sugar, which is itself a food with energy value. It adds to the daily intake of calories, an unhappy thought for people who want to control their weight, and helps to promote tooth decay. Sugar intake must also be avoided or limited by diabetics. For these and other reasons, artificial sweeteners are used in enormous amounts.

However, artificial sweeteners introduce other problems. *Cyclamates*, with 20 to 30 times the sweetening power of sugar, were banned in Britain and many other countries because they were found to be cancer-causing when fed to laboratory animals in large quantities. *Saccharin*, with more than 300 times the sweetening power of sugar, was banned in the United States by the Food and Drug Administration (FDA) for the same reason but was granted several reprieves by the US Congress. It is allowed in Britain. A newer sweetener, *aspartame*, has been linked to possible health problems and been the subject of long investigations. It is also expensive and in liquid form, as in soft drinks, loses some of its sweetening power.

Sorbital and *xylitol* are natural substances found in fruits, berries, mushrooms, and some other foods. Sorbitol (E420) is only half as sweet as sugar and is used in diabetic foods. In 1978, laboratory studies showed that xylitol increased the incidence of bladder tumours when fed to mice in large doses. The FDA in the United States moved to ban the sweetener but never took final action.

SYNCHROCYCLOTRON See PARTICLE ACCELERATOR

SYNFUELS See SYNTHETIC FUELS

SYNOVIAL FLUID See ARTHRITIS

SYNTHETIC FUELS (SYNFUELS)

Car fuel, jet fuel, and similar liquid fuels obtained from sources other than petroleum. The Arab oil boycott of 1973 and sharp rises in the price of oil focused attention on the need for alternative sources of fuel. Such sources have been developed, but in most cases the fuels produced cost more — sometimes very much more — than ordinary fuels. Some fuel sources include these:

Tar sands Also called *oil sands*, these are mixtures of sand and tarlike materials. They are mined and washed with hot water to extract the tar, then distilled to produce petrol and other liquid fuels.

Oil shale This is a type of limestone whose pores are filled with *kerogen*, a rubberlike material. When the limestone is mined and then heated, the kerogen produces a crude oil for distillation.

Liquefaction of coal The coal is heated under very high pressure in the presence of a catalyst. A petroleumlike liquid results; this is distilled, yielding liquid or gaseous fuel, as desired.

Coal, wood, agricultural wastes Chemical treatment of such organic materials, under high pressures, yields methanol (wood alcohol). Alcohol can be mixed with petrol (the mixture is called gasohol) for use in cars.

See also ANALYSIS; CATALYSIS AND CATALYSTS.

SYSTOLIC PRESSURE   *See* BLOOD PRESSURE

TAILPLANE

A fixed horizontal surface at the tail of an aeroplane or glider, used to maintain the up-and-down stability of the aircraft.
See also AERODYNAMICS.

TAPE RECORDING   *See* SOUND RECORDING

TAR SANDS   *See* SYNTHETIC FUELS

TAXONOMY   *See* BIOLOGY

TAY-SACHS DISEASE

A hereditary disease of the nervous system.
See also GENETIC DISEASES.

T CELLS   *See* IMMUNE SYSTEM

TECHNOLOGY   *See* SCIENTIFIC TERMS

TEMPERATURE-HUMIDITY INDEX (THI)

A number, also known as the comfort index (CI) or the discomfort index (DI), calculated from a formula that takes into account the temperature and relative humidity of the air. In general, a temperature of 20 to 21°C (68 to 72°F) and a relative humidity level of 55 to 60% is considered comfortable. Very humid air, especially at high temperature, is sticky and uncomfortable. The higher the temperature-humidity index, the less comfortable the air.
See also RELATIVE HUMIDITY.

TERRESTRIAL PLANETS   *See* SOLAR SYSTEM

TESTOSTERONE   *See* STEROIDS

TETRACYCLINES   *See* ANTIBIOTICS

TETRAHYDROAMINOACRIDINE
See ALZHEIMER'S DISEASE

TEVATRON See ELECTRON VOLT

THALASSAEMIA

A hereditary blood disorder, also known as *Cooley's anaemia*. See also GENETIC DISEASES.

THD See HIGH FIDELITY

THEORY, SCIENTIFIC See SCIENTIFIC TERMS

THERMAL PRINTER See PRINTER

THERMOCHEMISTRY See CHEMISTRY

THERMONUCLEAR REACTION See NUCLEAR ENERGY

THI See TEMPERATURE-HUMIDITY INDEX

THORIUM See RADIOACTIVITY

THUNDER AND LIGHTNING

Lightning comes in a variety of forms: streaks, sheets, bolts, and even balls; thunder, in peals, cracks, rumbles, grumbles, and growls. Yet these natural wonders all arise from the same basic causes, which can be examined over a carpet on a dry day.

In a darkened room, shuffle your feet across the carpet; then touch a metal radiator water pipe. Flash! minilightning; crackle! minithunder. Shuffling your feet scrapes a few billion electrons off the carpet and onto you. Electrons repel each other. Your approach to the radiator or pipe enables the electrons to stream off you (a stream of electrons is an electric current) onto the metal (a good conductor of electricity) and via this metal pathway to the earth's capacious bosom. The electrons, in leaping across, cause the air to glow, somewhat like what happens in a fluorescent tube. They also heat the air (to about 25,000°C/45,000°F in a real bolt). The heated air expands suddenly, producing the sound effect. Any sudden expansion of air can cause sound. Stick a pin in a balloon, and the compressed air escapes and expands, with a pop. Clap your hands, and you catch and compress a handful of air that escapes and expands with a crack.

Outdoors, sun-warmed air and water vapour molecules rise continuously from ground level to higher altitudes. During their shuffling colliding ascent they build up heaps of electrons — electric charges. Eventually, their mutual repulsion causes them to leap, with a push of 15 million volts or more, to a less charged or neutral area, then back again, over and over, in a series of oscillations, until they reach neutrality.

The oscillations are like what happens when you reach a high point on a swing and stop pushing: back and forth from a high point (maximum energy) to one somewhat less high, and still less high, until you reach the zero energy level.

The leaps of lightning may be from one part of a cloud to another part, or from one cloud to another, from cloud to earth, or vice versa. Big jumps (of 130 kilometres/80 miles or more) generally produce deep-throated thunder; smaller jumps make high-pitched cracks.

The shape of lightning bolts is determined mainly by the conductivity of the air. Electrons streak through the air along the path of least resistance (greatest conductivity), which may be zigzag, forked, or sinuous. If the lightning is entirely within a cloud, the shape of the bolt is diffused and obscured; this is *sheet lightning*. Sometimes a cloud acts as a reflector of lightning too distant to be heard; this is called *heat lightning*, because it usually occurs along the distant horizon at the end of a hot day. A rarely seen form is *ball lightning*, a small luminous ball that hovers above the ground for a few seconds, pops, and disappears. Its precise cause is unknown, but its German name is charming – *Kugelblitz*.

Safety rules about lightning (Notice that we didn't say 'and thunder'. Though frightening, thunder rarely threatens the eardrums; *see* DECIBEL.) As regards lightning, consider the following points:

- Avoid being under or near high trees. There are several reasons, one of them being that electric charges, when they achieve full voltage and start to flow, find the shortest path that will conduct them. A treetop is closer to the clouds than the ground itself.
- For the same reason, avoid being out in the open or exposed on high ground.
- Avoid touching metal fences, metal structures, and vehicles. Being good conductors and close to the ground, they invite electric charges; touching them includes you in the invitation.
- However (and this seems odd), the interior of a metal vehicle is a safe place to be, if you can't escape into a building. As previously mentioned, electrons repel each other. Electric charges striking a car repel each other and distribute themselves along the surface and away into the ground and air, but not inward, where they would be further crowding each other.
- Avoid handling flammable fuels such as petrol during a thunderstorm, for too obvious reasons.

A final note: On most electric power transmission lines you will see a single wire above the several cables that carry the power. This single line, connected to the top of each pylon, encourages lightning to flow between the air and the ground, thus bypassing the transmission lines. In the same way, metal lightning rods outside a building enable lightning to flow harmlessly between the air and the ground.

THYMINE   See DNA AND RNA

THYMUS GLAND   See IMMUNE SYSTEM

THYROID GLAND   See HORMONE

THYROXINE   See HORMONE

TIA   See STROKE

TIDAL ENERGY SYSTEMS

A method by which the ebb and flow of the tides is used to produce electricity.
See also ENERGY RESOURCES.

TIME, RELATIVE   See RELATIVITY

TISSUE CULTURE

A sophisticated method for growing living tissue *in vitro* (Latin, in glass) or *in vivo* (in what is alive).

For *in vitro* culture, tissue from a plant or animal is placed in a medium, usually liquid, in which an assortment of substances has been dissolved. Some of these nourish the tissue; others approximate the chemical makeup of its normal environment. Contamination by bacteria, a major obstacle in the early days of tissue culture, is controlled by ANTIBIOTICS added to the medium.

In vivo tissue culture is carried out by implanting tissue into a living plant or animal. For example, a scientist might transfer cancerous tissue from a mouse to a number of other mice, to test the effects of varying concentrations of an anticancer drug.

Both the *in vitro* and *in vivo* methods have been refined to a point where a single cell isolated from a tissue can be cultured successfully. The technique is called *cell culture*.

Virtually every advance in biology and medicine in the last half-century owes something to the techniques of tissue culture and cell culture. For example:

- Vaccines against viral diseases such as polio. The virus must be cultivated on living tissue *in vitro*.
- Recognition under the microscope of chromosomes that denote various kinds of disorders in the fetus (such as DOWN'S SYNDROME). The fetal tissue for examination is first grown *in vitro*.
- Growing skin for grafting onto extensive burns.
    ici1.6l● Cancer and immunology research.
- Propagation of desirable varieties of asparagus, orchids, and many other plants, using extremely thin slices of tissue; identical plants (clones) by the thousands are produced in this way.

TISSUE PLASMINOGEN ACTIVATOR
See GENETIC ENGINEERING

TISSUE REJECTION See IMMUNE SYSTEM

TOKAMAK   See NUCLEAR ENERGY

TONE CONTROL

An electric circuit in a record player or radio that controls the deepness (low pitch) or shrillness (high pitch) of a sound. Pitch depends on frequency (the rate at which a sounding object vibrates). For example, the middle C string of a piano, when struck, vibrates 264 times per second, producing a sound called the *fundamental tone* of the string. Most sounds, however, result from a mixture of frequencies.

The string also vibrates at the same time in fractions of its full length (in halves, thirds, etc.) at higher frequencies, producing higher-pitched sounds called *overtones*. These blend with the fundamental tone to give what we regard as the special piano sound. A violin string, playing the same note of C, produces a different mixture of overtones that blend with the fundamental. The blending is the sound we recognize as the special violin sound. Each kind of musical instrument produces a characteristic mixture that identifies it as oboe, flute, French horn, and so on. In the case of the human voice, the blend of tones even enables us to recognize the speaker.

Ideally, the sound of an instrument or a voice should emerge from the speakers of a musical system exactly as it entered at the recording studio. But it doesn't (*see* HIGH FIDELITY). The mix – the proportion of fundamentals and overtones — is altered by all kinds of forces. To repair the damage and to modify the sounds to suit the listener's preferences, tone control circuits are built into the system.

Inexpensive tone controls, of the sort found on low-grade systems, simply clip away the highest overtones. This lawnmowerlike method leaves a sound that is muffled and woolly, lacking the sharp edge that is the signature of various instruments. On higher-quality musical systems, the tone control is actually two or more circuits. One circuit, operated by the bass control knob, emphasizes the fundamental notes. The treble control circuit does likewise for the higher-pitched overtones. Still more elaborate systems have several tone controls, each controlling the emphasis of a specific range of fundamentals or overtones.

TORNADO

A dark, funnel-shaped, extremely intense whirlwind, formed during a severe thunderstorm. Its darkness is due to the presence of water vapour, dust, and debris within. Although the term is derived from Latin *tonare*, 'thunder', a tornado is far more violent and destructive than its parent thunderstorm. Example: in Topeka, Kansas, a heavy upholstered sofa was plucked out of a hallway, blown 800 or so metres (½ mile), and deposited on the roof of a three-storey office building. The air within a tornado is under negative pressure: that is, at much lower pressure than the surrounding air. Thus if a tornado surrounds a house, especially one with doors and windows closed, the house literally explodes.

A tornado's lower end, where it does maximum damage, is fortunately rather narrow, perhaps a couple of metres to 1.5 kilometres (about a mile) in diameter. A tornado over water is called a *waterspout* because the negative pressure draws the water into the tip of the funnel.

Accompanying the destructive negative pressures of a tornado there is also damage done by huge circular winds, lightning, and heavy rainfall.

The exact conditions that produce tornadoes are not clearly understood, but they include the presence of warm, moist air at lower levels and cold, dry air and high winds at higher levels. (This situation often occurs along cold fronts.) The high-speed rotary motion is similar to the spin of water emptying out of a bath as a large volume of water crowds out through a relatively small opening. In a tornado, a large volume of air flows inwards towards an ascending column of sun-heated, expanding, low-pressure air.

TOTAL ECLIPSE *See* ECLIPSE

TOTAL HARMONIC DISTORTION See HIGH FIDELITY

TOXICOLOGY See MOLECULAR BIOLOGY

TOXIC-WASTE DISPOSAL

In 1978 Niagara Falls, New York, famous for honeymooning, became the centre of a horror story. About 25 years earlier, a chemical manufacturing company had disposed of 20,000 tonnes of highly poisonous wastes by burying them in a landfill at the site of an old canal (named the Love Canal, no less). Now women living in the area were experiencing miscarriage and stillbirth rates higher than normal as a result of exposure to the buried chemicals. Ten years later, more than $200 million had been spent in relocating about 1000 families and in cleaning up, a job that has not yet finished.

The Love Canal is only one of thousands of places where the toxic wastes of industry have been stored or dumped. Distance from a dumping site is no guarantee of safety, because rainwater seeping through the dump carries contaminating material through the soil and may reach wells or rivers serving as water supplies (see GROUNDWATER). An alternative to landfill disposal is the burning of toxic compounds in specially designed incinerators. Much of this is carried out at sea — in the North Sea over 100,000 tonnes of toxic waste is disposed of each year. Now, scientists are finding high levels of extremely poisonous chemicals in sediments on the seabed, resulting from incomplete burning. A global ban on all incineration at sea by 1995 has been proposed.

For the future, better alternatives are available:

Build better dumps 'Secure' landfills have plastic liners, earth covers, and mechanisms for recapturing any liquid that leaks out. Though better than ordinary landfills, they merely postpone the problem of ultimate disposal.

Break down the poisons Many toxic chemicals can be changed to simpler, nontoxic compounds by very hot water under great pressure, by strictly controlled incineration, or by neutralization with other chemicals. Research is also being carried out into the possibility of producing genetically engineered strains of bacteria which will degrade toxic chemicals.

Don't produce the poisons in the first place Some manufacturers have devised formulas and processes that reduce or eliminate the output of toxic wastes. This is clearly the best solution.

TPA See GENETIC ENGINEERING

TRAILING EDGE See LEADING EDGE

TRANSCRIPTION OF DNA See DNA AND RNA

TRANSDUCER

A device that converts (Latin *transducere*, to lead across) one form of energy into another. For example, a toaster converts electrical energy into heat energy.

All objects in action can be classed in two ways: as performers of function and as transducers. An electric cooker's function is to cook food, which it does by acting as a transducer of electrical energy into heat energy. A radio receiver's function — to reproduce sound at a distance, without wires — is accomplished by transducing electromagnetic energy (in radio waves) into the special kind of mechanical energy called sound. A transducer can also convert one aspect of a form of energy into another aspect. For example, a hand eggbeater converts the rotary motion of a hand into the rotary motion of the beater blades. Both are aspects of mechanical energy.

Scientists and technicians generally reserve the term *transducer* for more specifically scientific devices. A microphone is a transducer because it converts sound into electrical energy; a solar cell qualifies because it changes light into electrical energy. But the dictionary leaves the field open to all 'leaders across' of energy. A breeze rattles the window blind (mechanical energy into sound); an electric buzz saw rips through a plank (electrical to mechanical, with the by-products of heat and sound). Also in the category of transducer belong a door knocker, an electric doorbell, a television receiver, and a fire-engine siren.

TRANSFER RNA See DNA AND RNA

TRANSFORMER

An astonishingly simple, efficient, and useful device for changing the voltage of an alternating current. An example is the common bell-ringing transformer, which lowers 240-volt house current to about 6 volts, enough to operate an ordinary house bell or chime. Its voltage-reducing action classes it as a *step-down transformer*. Your television receiver contains an example of a *step-up transformer*. It raises the house current voltage to 20,000 or more to operate the picture tube. Other transformers in your TV receiver provide various voltages for lighting dials, powering loudspeakers, and so on.

Some of the most commonly seen transformers are the small substations or *transformer chambers* situated every 400 metres (¼ mile) or so in residential and industrial areas. They step down electrical currents of 33,000 or 66,000 volts from the main substation to 415 or 240 volts for use in homes, factories, and office buildings.

A transformer has three basic parts: an iron core and two coils of copper wire, the primary and secondary coils. Alternating current (AC) flows into the primary coil. AC flows first in one direction, then in the other. The standard rate of reversal in Britain is 50 back-and-forths, or cycles, per second (called 50 hertz or 50 Hz). Each change – forwards, stop, reverse, stop, forwards again – causes a magnetic field to build up around the primary coil and then collapse. The buildups and collapses sweep across the iron core, into the secondary coil, and back. The sweeps propel electrons in the secondary coil back and forth in the same rhythm, setting up a rapidly reversing 50-Hz secondary AC. This process of inducing electrons to move by means of magnetism is called *electromagnetic induction*. The voltage of the *induced current* is in direct ratio to the number of turns of wire in the two coils. For example, to step down 240 volts of house current for use with a 6-volt bell, the primary coil could be wound with 240 turns and the secondary coil with 6 turns, and *voilà!* Any ratio of 240 to 6 will do the job: 480 to 12, 720 to 18, and so on. The greater the number of turns on both coils, the heavier-duty job the transformer can do.

In a step-up transformer, the primary coil has fewer turns than the secondary. The transformer in a TV set, for example, could have a ratio of 240 turns in the primary to every 20,000 in the secondary.

Why aren't the power-line transformers eliminated by simply generating electricity at lower voltages – 240, for example? Because every current loses some energy as it flows, mostly in the form of heat that wastefully warms the wires and their surroundings. This loss occurs with all electric currents, but it is proportionately less when the voltage is very high.

Working on AC, the quiet, efficient transformer is one of the most useful of electrical devices. That's why most of the world's electrical needs are served by AC. Direct current (DC), which flows in one direction without reversals, doesn't work in transformers, because it

TRANSFORMER

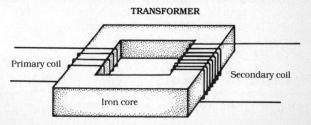

doesn't produce repeated rising and collapsing magnetic fields, as AC does.

TRANSFORMER CHAMBER See TRANSFORMER

TRANSFORM FAULT See PLATE TECTONICS

TRANSGENIC ORGANISMS See GENETIC ENGINEERING

TRANSIENT ISCHAEMIC ATTACK See STROKE

TRANSISTOR See SEMICONDUCTORS

TRANSIT OF PLANETS See OCCULTATION

TRANSMISSION ELECTRON MICROSCOPE See MICROSCOPE

TRANSPIRATION

The passage of vapour or gas through pores or membranes. In plants the oxygen, carbon dioxide, and water vapour associated with carbohydrate manufacture pass through *stomata* (singular *stoma*). These are tiny openings in the leaf, usually on its underside. The size of the opening is regulated by changes in the shape of a pair of *guard cells* that surround it.

TRIGLYCERIDES See CHOLESTEROL

TRIIODOTHYRONINE See HORMONES

TRIPLET CODE See DNA AND RNA

TRITIUM See NUCLEAR ENERGY

TROPICAL CONTINENTAL AND MARITIME AIR MASSES See AIR MASS ANALYSIS

TROPICAL CYCLONE See HURRICANE

TUMOUR See CANCER

TWEETER See LOUDSPEAKER

TYPHOON See HURRICANE

ULTRASONICS

The science of sound beyond the range of audible sound, named from the Latin for 'beyond sound'. Human hearing extends to sounds up to 20,000 hertz per second (cycles). Beyond that are high-pitched sounds audible to some animals; for example, dogs respond to dog whistles pitched at 25,000 Hz, inaudible to human whistle blowers, and bats emit squeaks at 100,000 Hz and guide themselves in the dark by listening for their echoes from nearby obstructions (see SONAR).

There are many interesting artificial ways of generating and using ultrasonics. *Ultrasonic generators* are like record player tone arms in reverse. In a tone arm, the mechanical vibrations of a wiggling needle are converted into electrical vibrations. In an ultrasonic generator, electrical vibrations are converted into mechanical vibrations. Usually this is done by passing a high frequency (many vibrations per second) alternating current through a crystal, such as quartz. This causes the crystal to vibrate in step with the current. Such energy conversions — mechanical vibrations to electrical or vice versa — are examples of the *piezoelectric effect* (Greek *piezo*, press, squeeze). The cystal's vibrations cause the air to vibrate and produce sound (inaudible to you because the frequency is higher than 20,000 Hz).

The uses of ultrasonics are many, ranging from cleaning silverware to catching burglars to drawing electronic portraits (*sonograms*) of unborn babies. Ultrasonic devices work by either delivering focused energy or detecting and measuring vibrations from an ultrasonic receiver.

Delivering focused energy Ultrasounds can be aimed, focused, and reflected almost like light beams. Specific frequencies can cause certain kinds of molecules to vibrate while others are left unmoved (see RESONANCE). One specific frequency can cause solder to vibrate, heat up, and melt while only warming the metal being soldered. Another frequency loosens plaque from teeth but not teeth themselves. Another frequency causes kidney stones to be pulverized without affecting the kidney itself.

Detecting and measuring ultrasounds Ultrasonics can be used for guarding banks, offices, and factories. An ultrasonic beam is aimed and reflected so that it criss-crosses a room several times and then strikes a detector, a kind of ultrasonic microphone. An intruder walking into the path of the sound sets off a remote signal, if the intruder is large enough to block off sufficient sound energy. A human intruder will; a cat or mouse won't. Another kind of detecting and measuring is done by using ultrasound as a kind of X-ray, without the risks of X-ray exposure. A beam of ultrasound travels directly through a homogeneous substance, but if it reaches a different substance (at an interface), the beam is reflected and forms a sonogram. In this way, ultrasonic detectors can locate cracks or bubbles in metal castings. Similarly, and much more important, interior organs of the body and fetuses can be located and outlined. For example, an echocardiograph machine clearly shows the opening and closing of the valves of the beating heart.

ULTRAVIOLET RAYS See RADIATION ENERGY

UMBRA See ECLIPSE

UNCERTAINTY PRINCIPLE

A principle stating that 'the position and momentum of an atomic particle cannot both be known accurately at the same time'. On the surface, hardly a world-shaking matter, yet the uncertainty principle was a shock to most scientists when it was first announced in 1927 by the German physicist Werner Heisenberg (1901–1976). Today the principle continues to dominate scientific philosophy and experimentation. The cosmologists who figured out the big bang theory and the physicists who devised the picture tube of your TV had to take into account the uncertainty principle because both fields involve the motion of atomic particles: electrons, protons, and neutrons. In fact, any work with atomic particles must deal with the uncertainty principle.

To understand the importance of this principle, consider the pre-Heisenberg era, when atoms and their particles were considered to be like tiny balls whose behaviour was as certain and as predictable as that of billiard balls. Start a ball rolling on a billiard table. With enough facts, and perhaps the help of a computer, a physicist can predict the exact path of the ball, with every collision and rebound, until it stops rolling. Among the 'enough facts' there would have to be the ball's weight, its speed, and the direction of its motion (these three factors together determine its momentum), air resistance, friction, and elasticity of the ball and the table cushion.

How comforting – not only to billiard players and physicists (pre-Heisenberg) but to all people attempting to draw a plausible picture of the real world. Perhaps the ultimate picture of certainty was drawn by the French astronomer Pierre-Simon de Laplace (1749–1827). Laplace suggested that if at a certain moment the position and momentum of every particle in the universe were known, the entire history of the universe, past and future, could be determined from that moment, forwards or backwards.

But alas for Laplace, the universe is not a game of billiards. True, billiard balls, planets, and other such tangible lumps of matter do behave predictably, but when we get down to the ultimate small stuff — the particles of particles (electrons, quarks, etc.) — we find not exact predictability but only statistically *probable* behaviour. Yes, if we hurl a million electrons in sequence at a target, with the same momentum, they will arrive at predictably approximate positions, clustered at and around the centre, like an expert rifleman's target. *Approximate* positions, not identical. Furthermore, as Heisenberg stated, we can never measure both momentum and position of a particle with accuracy; the closer we come to one, the fuzzier the other becomes.

Nobody is quite comfortable with the uncertainty principle or the statistical picture of matter, although experiment after experiment bears out the correctness of the principle. On the subatomic scale, position and momentum are uncertain quantities. We can never pin them both down at once. Why not? What causes the uncertainty? And can we speak of the *cause* of an uncertainty? Physicists and philosophers have a field day with such questions.

Another example of subatomic unpredictability is radioactivity in relation to time. On the ordinary (large) scale, time can be measured with certainty and exactness by planetary revolutions and all sorts of clocks. Consider one example of an exact and certain behaviour of large-scale matter as it relates to time: a fireworks manufacturer builds 1000 identical rockets, each with a time-delay fuse set to go off exactly one hour after being triggered. If all are triggered simultaneously, all should go off simultaneously — *and they do*. But at the subatomic level, time turns statistical *average* on us, and we encounter an uncertainty-of-time situation.

For example, consider the element carbon, which has a special form, or isotope, carbon 14 (C-14). An atom of C-14 may emit a subparticle and turn into an atom of nitrogen. It simultaneously emits a tiny flash of light, which can be detected. We find that the atoms of C-14 don't all flash at once. One atom emits a flash here and another emits a flash there, and that's it for those two atoms. And which one will be next? Nobody knows, but this *is* known: after 5730 years, half of the atoms will have emitted their single flash. This length of time is called the *half-life* of C-14. After another 5730 years, half of the remaining half will have flashed, and so on every 5730 years. But which ones will they be, and why those?

Statistically, you could use a block of carbon as a kind of long-term clock (and indeed scientists do this in a process called *radiocarbon dating*), but any individual atom of C-14 is totally unreliable as a measurer of time. *When* will a particular atomic particle perform its little act? *Where* will it be at the end of the performance? We can never be certain of both answers together. That's Heisenberg's uncertainty principle.

UNIVERSAL PRODUCT CODE *See* UPC

UNSATURATED FATS *See* CHOLESTEROL

UNSTABLE NUCLEI *See* RADIOACTIVITY

UPC

Abbreviation of *u*niversal *p*roduct *c*ode; also called *bar code*. A system for labelling a package or product with a code based on a pattern of lines and bars. For example, note the UPC on a box of breakfast cereal. The pairs of thin lines at the right and left are the code for 'These are the ends of the pattern'. The cluster of lines and bars from the left to the middle is the code for the manufacturer. From the middle to right, the code indicates that this is a 500-gram box of bran flakes. The pair of thin lines in the centre separates the two halves of the code.

Imagine that the lines were printed with raised ink. If you drew your fingernail across the pattern often enough, you could learn to indentify it by the succession of thick and thin lines and spaces. At the checkout counter of a UPC-equipped shop, the 'fingernail' is a very thin, lowpowered laser beam, shining out of a device called a scanner.

The laser beam sweeps across the pattern of lines and spaces, which may be thick, medium, or thin. The beam is reflected unevenly because the white or light-coloured spaces are better reflectors than the dark lines. A photoelectric cell, or electric eye, in the scanner receives the reflections — some bright, some dim — and converts them to electric pulses, like dots and dashes. The pulses go into a computer that 'recognizes' the pattern and does several jobs in a tiny fraction of a second.

1. It sends a command to the cash register to display and print on tape the name of the item and its price.
2. If the box has been presented to the scanner upside down, it inverts the reading to compensate.
3. It directs its memory to deduct one 500-gram box of bran flakes from the shop's inventory.
4. When the inventory of this item falls below a certain number, the computer alerts the manager to order more.
5. It can deal with linked prices, such as two purchases for £1.99. The machine charges £1.00 for the first box and 99p for the second, even if several intervening items have been charged.

URACIL   See DNA AND RNA

URANIUM, DECAY OF   See RADIOACTIVITY

URANIUM 235   See NUCLEAR ENERGY

URANIUM 238   See NUCLEAR ENERGY; RADIOACTIVITY

UREA   See PROTEINS AND LIFE

VACCINATION See IMMUNE SYSTEM

VALINE See DNA AND RNA

VAN ALLEN BELTS

Belts of electrically charged particles, trapped high above the earth by its magnetic field, or *magnetosphere*. The two belts, which are about 1000 to 5000 kilometres (600 to 3000 miles) and 15,000 to 25,000 kilometres (9300 to 15,500 miles) above the earth, extend approximately 64,000 kilometres (40,000 miles) from earth in the direction of the sun but may extend over 1.5 million kilometres into space in the opposite direction. The charged particles, which come from the sun (*see* SOLAR WIND), are the cause of AURORAS and interfere with radio and TV reception.

The discovery of the belts was one of the many accomplishments of the INTERNATIONAL GEOPHYSICAL YEAR, 1957–58. The belts were detected by sensors aboard *Explorer I*, the first US satellite, in an experiment conducted by Dr James A. Van Allen, the American physicist after whom the belts are named.

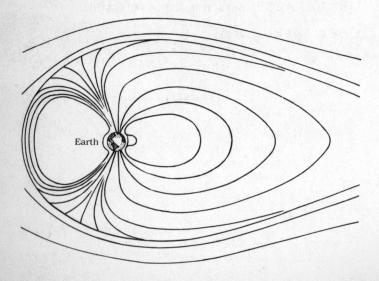

VARIABLE See SCIENTIFIC TERMS

VEIN See CIRCULATORY SYSTEM

VENTRICLES OF THE HEART
See CARDIAC PACEMAKER

VENULE See CIRCULATORY SYSTEM

VICTORIA FALLS, FORMATION OF See FAULT

VIDEOTAPE

A system for storing visual information, such as television programmes, on magnetic tape. There are no pictures on a recorded videotape, only clusters of magnetized and unmagnetized metal particles. The magnetized clusters represent the signal 'yes'; the unmagnetized clusters represent the signal 'no'. Anything that can be coded into a series of yes and no signals can be recorded on magnetic tape. Here is an example of how a very simple black-and-white picture can be coded as shown.

Read every white square as 'yes' and every black square as 'no'.

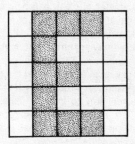

Read across (scan) from left to right, top to bottom. The first line would read 'yes, no, no, yes'. The second line is 'yes, no, yes, yes, yes'. Continue for a total of five lines, and you have coded the letter *E*. Now suppose that instead of your eyes and voice, you used an 'electric eye' and an electromagnet. A piece of blank magnetic tape is pulled steadily across the electromagnet. Each yes would produce a magnetized cluster, and each no would leave an unmagnetized one. This recorded tape could then be pulled across a mechanism that senses the magnetized and unmagnetized sections, the yeses and noes. Each yes sends a pulse of electric current that makes a dot of light on a TV tube. Each no sends nothing, leaving a dark area. Line by line the letter *E* is reproduced.

Real videotape and TV differ from this extremely oversimplified model in many ways, mainly the following:

1. The model uses five horizontal rows and five vertical columns for a total of 25 squares, or elements. Real television uses 625 rows and 833 'columns' for a total of 520,833 elements — obviously, a much finer-detailed picture.
2. In colour television the face of the picture tube is composed of groups of dots, called pixels (picture elements). Each pixel consists of three dots, one for each of the primary colours of light: red, green, and blue. A dot, when struck by a yes signal (an electron beam), gives a momentary glow in its colour. All three colours glowing at full brightness add up to white. Reducing the brightness of one or more colours changes the mixture, whose actual colour depends on the brightness of each constituent colour.

VIRUSES

The smallest, simplest living things, and the cause of some of our deadliest diseases. A big virus may be 0.00025 millimetre ($1/100,000$ inch) in diameter, a small one perhaps one-tenth that size. Viruses are rod-shaped or spherical bits of nucleid acid (See DNA AND RNA) wrapped in a coat of protein.

DNA is the hereditary material that makes up the *genes*, the 'instructions' that determine the traits of every living thing, whether it develops as an oak or an octopus; how it looks; and what chemical reactions it can perform. A virus relying on its own meagre stock of genes can't obtain energy to grow or carry on any other process needed for life. Is it alive then? That question has started many a juicy debate; we'll point out only that viruses persist from one generation to another, as if alive. How do they manage that?

Consider a *bacterial virus* — one that attacks bacteria (also called a *bacteriophage*, from Greek *phagein*, to eat). A: The virus's DNA is injected through its 'tail' into the bacterium, leaving the protein outside. B: Minutes later a new virus is seen within the bacterium, then another and another. C: In less than half an hour the packed bacterium breaks open and hundreds of viruses escape, ready to attack more bacteria.

What happened in that short time? Normally, the genes of a bacterium control its usual functions: the making of proteins, more DNA, cell walls, and so on. When the marauding viral genes break in, they reprogramme the cell's machinery to make *viral* DNA and *viral*

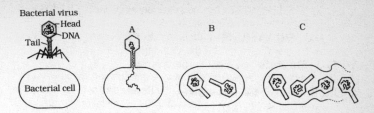

protein coats. Thus the parasitic virus, which is helpless when outside a living cell, enslaves its hosts — not only bacteria but also plants, animals, and humans. Here are some examples:

In plants Viruses cause *mosaic diseases* in many crops of economic importance: tobacco, tomatoes, soyabeans, and others.

In animals Cattle foot-and-mouth disease, canine distemper, some cancers, rabies, and other diseases are caused by viruses.

In humans AIDS, at least one form of cancer, herpes, chicken pox, colds, flu, polio, hepatitis, and others are viral diseases.

The body's IMMUNE SYSTEM responds to virus attacks in two ways:

1. It produces *antibodies* that inactivate the virus. It also 'remembers' how to produce armies of the antibodies, in case there is a later attack by the same virus. Doctors often use dead or weakened viruses in vaccination to prevent some virus diseases — polio, for example. A dead or weakened virus can't cause a real illness, yet it elicits the same *immune response*, the production of antibodies.
2. It produces INTERFERON, a protein that prevents the virus (and other kinds of viruses as well) from spreading to healthy cells.

Antibiotics, which are spectacularly useful in fighting diseases caused by bacteria, are virtually useless against viral diseases. Pathogenic (disease-causing) bacteria in the blood or among the body cells do their damage mainly by producing toxins (poisons) that affect the normal functioning of the body. Antibiotics control these bacteria by inhibiting their growth, or by interfering with their ability to form the cell walls that keep the contents of the bacterial cells intact.

Unlike bacteria, pathogenic viruses are not cells. They are bits of nucleic acid able to reproduce *only within body cells*, by subverting the cells' own nucleic acids. Thus drugs capable of damaging viruses can damage the body cells as well.

See also DNA AND RNA; PROTEINS AND LIFE.

VISIBLE SPECTRUM *See* RADIANT ENERGY

VISUAL BINARIES *See* BINARY STAR

VITAL SIGNS *See* CIRCULATORY SYSTEM

VOICE COIL *See* LOUDSPEAKER

VOICE TRANSMISSION, LASERS IN *See* LASER

VOLCANO *See* ROCK CLASSIFICATION

VOLT *See* ELECTRICAL UNITS

VOLTAGE *See* ELECTRICAL UNITS

VOLUME CONTROL *See* LOUDNESS CONTROL

WARM FRONT See AIR MASS ANALYSIS

WATCHES, QUARTZ See QUARTZ

WATER CYCLE

The movement of water from clouds, condensing and precipitating to earth in the form of rain or snow; its temporary stay in the form of mist, vapour, and groundwater; and its return to form clouds again. That last part, the return, was a source of bewilderment to history's most versatile genius, Leonardo da Vinci (1452–1519). In his treatise on water, Leonardo speculated on how rivers could begin on mountaintops. Perhaps, he conjectured, ocean water flows through deep horizontal tunnels to the bases of mountains; then some kind of 'natural engines' freshen the seawater and elevate it through vertical tunnels to the tops of mountains.

Here is a brief outline of the commonly accepted modern theory of the water cycle:

Water exists in 3 ½ forms: (1) solid, as ice or snow; (2) liquid, as water; (3) gas, as invisible water vapour; (3 ½) as a transitional state, visible in tiny droplets as fog, mist, clouds, and the stuff emerging from boiling water in kettles.

The proper name of the water cycle is the *hydrologic cycle* (Greek *hydor*, water). It is the circulation of the waters of the earth between land, seas, oceans, and atmosphere. Since circulation implies no beginning and no end, we can get on anywhere, so we choose to begin at the place where we pooh-poohed Leonardo:

Energized by the heat of sunlight, water evaporates from the oceans into the atmosphere (see the illustration on the next page). It may remain for a while as invisible water vapour but eventually condenses to form clouds. Much of this cloud water precipitates as rain; however, under certain conditions of low temperature, it may evaporate and condense into snowflakes. (Snow is not frozen rain, but sleet is.) Most rain and snow falls back into the ocean for the simple reason that most of the earth's surface is ocean.

WATERFALL FORMATION

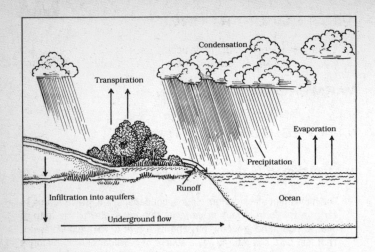

Water falling on land follows several routes:

1. Some water runs into underground layers of rock and soil (aquifers), where it is held as GROUNDWATER.
2. Some water is taken in by green plants and serves as the basic raw material, together with carbon dioxide from the air, in the incredible process of photosynthesis, by which plants make food. Excess water evaporates from the leaves by the process called TRANSPIRATION.
3. Some rain falling above land and sea evaporates directly back into the atmosphere, without having experienced life on earth, but, this being a cycle, there's always a next time.

About 97% of the earth's water is in oceans — salt water — but during the process of evaporation, the water alone moves up into the atmosphere. Of the remaining 3% — unsalty — three-quarters is in solid form, as glaciers, and one-quarter is in lakes, rivers, inland seas, and in transition in the atmosphere.

WATERFALL FORMATION See FAULT

WATERSPOUT See TORNADO

WATER TABLE See GROUNDWATER

WATT See ELECTRICAL UNITS

WEAK NUCLEAR FORCE

One of the four known fundamental forces of the universe: also known as the weak interaction. Its influence is limited to the atomic nucleus.
See also GRAND UNIFIED THEORY.

WEATHERING *See* ROCK CLASSIFICATION

WEIGHT *See* MASS

WELL, WATER *See* GROUNDWATER

WHISKERS *See* FLUTTER

WHITE DWARF STAR *See* STELLAR EVOLUTION

WHITE ROT FUNGUS *See* ENZYME

WIND-ELECTRIC SYSTEM

A method designed to use the energy in winds for producing electricity.
See also ENERGY RESOURCES.

WINGS

In aerodynamics, a pair of bladelike structures fixed to the sides of aeroplanes or gliders. Their movement through the air produces the force that lifts the craft (*see* AEROFOIL). In a helicopter the large rotating blades perform the lifting function. Early aeroplanes had two pairs of wings (biplanes) and even three pairs (triplanes).
See also AERODYNAMICS.

WIRE DRAWING *See* MANUFACTURING PROCESSES

WOOFER *See* LOUDSPEAKER

WOW *See* FLUTTER

W − PARTICLE *See* GRAND UNIFIED THEORY

W + PARTICLE *See* GRAND UNIFIED THEORY

X-RAYS *See* RADIANT ENERGY

XYLITOL *See* SWEETENING AGENTS

ZOSTER *See* HERPESVIRUS DISEASES

Z° PARTICLE *See* GRAND UNIFIED THEORY

About the Authors

Herman Schneider, award-winning author of numerous science books and magazine articles (*American Home, Life, The Instructor, Highlights, Parents*), has taught at the City College of New York and was Science Supervisor for the New York City elementary schools. He has contributed to the production of numerous science films and television programmes.

Leo Schneider, also an award-winning author of science books, was a science curriculum writer for the New York City Board of Education and Senior Science Editor for the *New Book of Knowledge* encyclopedia.